Utilizing international land use regimes
to shape domestic forest policies in Indonesia

Muhammad Alif Kaimuddin Sahide

Utilizing international land use regimes to shape domestic forest policies in Indonesia

Dissertation

Submitted in fulfilment of requirements for admission to the degree of Dr. forest. at Faculty of Forest Sciences and Forest Ecology Georg-August University, Göttingen

Georg-August Universität Göttingen
2016

Bibliografische Information der Deutschen Nationalbibliothek
Die Deutsche Nationalbibliothek verzeichnet diese Publikation in der Deutschen Nationalbibliografie; detaillierte bibliografische Daten sind im Internet über http://dnb.d-nb.de abrufbar.
1. Aufl. - Göttingen: Cuvillier, 2016
Zugl.: Göttingen, Univ., Diss., 2016

Nonnenstieg 8, 37075 Göttingen
Telefon: 0551-54724-0
Telefax: 0551-54724-21
www.cuvillier.de

1. Auflage, 2016
Gedruckt auf umweltfreundlichem, säurefreiem Papier aus nachhaltiger Forstwirtschaft.

ISBN 978-3-7369-9216-0
eISBN 978-3-7369-8216-1

Dedicated to my family

Chair of Forest and Nature Conservation Policy
Faculty of Forest Sciences and Forest Ecology
Georg-August University Göttingen
GERMANY

This research is conducted at the working group "International Forest Regimes" at Chair Group of Forest and Nature Conservation Policy, Göttingen University, Germany. This research is a part of the comparative research on "Fragmentation of the International Forest Regime-Complex" conducted within the Working Group International Forest Regimes lead by Dr. Lukas Giessen.

This research presented in this book is funded by Indonesia Directorate General of Higher Education (DIKTI), German Research Foundation (DFG), and the EVA-MAYR STIHL Foundation.

Supervision: 1. Professor Dr. Max Krott
2. Lukas Giessen, Ph.D

Case Study and Country:

1. Muhammad Alif Kaimuddin Sahide (Indonesia); Utilizing international land use regimes to shape domestic forest policies in Indonesia; Göttingen
2. Agung Wibowo (Indonesia); Implementation of the international forest regime by bureaucratic politics in Indonesia; Göttingen
3. Sarah Lilian Burns (Argentina); The influence of the international forest regime in Argentina; Göttingen
4. Scott Cettie (USA); Discourse analysis in the international forest policy arena: discourse coalitions, discursive elites and discursive agency; Freiburg
5. Anja Eikermann (Germany); International law on biodiversity and its fragmentation; Göttingen
6. Almut Jenke (Germany); Forest inventories, reporting of forest information and their contribution to the policy process in decisions on sustainable forest management; Göttingen
7. Johannes Jürging (Germany); European forest governance as part of the EU's sustainability law; Göttingen
8. Sina Leipold (Germany); managing a fragmented field? Discourse elites at the crossroads of domestic and global forest policy in the USA, Germany and Sweden; Freiburg
9. Jacqueline Logmani (Germany); Implementation of the international forest regime by bureaucratic politics in Germany, Sweden and Poland; Göttingen
10. Md. Saifur Rahman (Bangladesh); The Bureaucratic Implementation of the International Forest Regime in Bangladesh; Göttingen

Declaration of originality and certificate of authorship

I, Muhammad Alif Kaimuddin Sahide, hereby declare that I am the sole author of this dissertation entitled: *Utilizing international land use regimes to shape domestic forest policies in Indonesia*. All references and data sources that were used in the dissertation have been appropriately acknowledged. I furthermore declare that this work has not been submitted elsewhere in any form as part of another dissertation procedure.

Göttingen, February 2016

Muhammad Alif Kaimuddin Sahide

Acknowledgements

I would firstly like to thank Professor Max Krott who was a great support and mentor throughout much of this research. I am grateful to professor Krott for guiding me to the way of critical thinking and objectivity and providing me the opportunity to continue my academia aspiration. A bunch of thanks is also given to Dr. Lukas Giessen, who gave me a lot of help and encouragement in the early stages of the study, who provided me a very high intensive supervision and his fruitful leader on the International Forest Regime working group. I am particularly thankful to Prof. Dr. Anja Jetschke who acted as the co-referee and a member of the examination committee, along with Prof. Dr. Renate Bürger Arndt.

Appreciation also goes to the colleagues in the International Forest Regime Working Group, with whom I was involved in fruitful discussions and collaboration. They are Jacqueline Logmany (Germany), Sarah Lilian Burns (Argentina), Agung Wibowo (Indonesia), and Md. Saifur Rahman (Bangladesh). Other colleagues in the Chair Group of Forest and Nature Conservation Policy, e.g., Kenji Nagasaka (Japan), Budi Dharmawan (Indonesia), Janina Heim (Germany), Ngo Duy Bach (Vietnam), and Do Thi Huong (Viet Nam) also helped create an atmosphere conducive to scientific endeavour. I thank all of them.

I would like to sincerely thank Indonesian Government Scholarship (DIKTI) for providing me 7-semester scholarship to undertake the doctoral study. I would also thanks to German Research Foundation (DFG) as well as the Eva Mayr-Stihl Foundation for providing financial support on my fieldwork

I would also sincere my thanks to my home based University in Indonesia, namely, the Dean of the Faculty of Forestry, and the Rector of Hasanuddin University in Indonesia. I very much thank and acknowledge their support.

I am particularly grateful to Dr. Dodik Nurrochmat for his valuable support during my fieldwork in Indonesia and Philippine. The fieldwork itself dealt with numerous ministries, non-governmental institutions, international organizations in national level (Indonesia) and Provincial level (South Sulawesi Province) as well as individuals, particularly ASEAN members' representative. I am particularly grateful to forest officials at the Centre of Foreign Cooperation of Ministry of Forestry for sincere appreciation.

Living away from family in home country is not easy. I am particularly thankful to all Indonesian students and families in Göttingen. Support from them cannot be discounted.

Finally, words might not be sufficient to express thanks to my parents. Without their extraordinary efforts and beliefs, I might not have reached this advanced education. Strength and inspiration were unlimitedly provided by my lovely wife, Irmayanti and two lovely sons, El Hurr Nur Muhammadi and Sajjad Makkarennu Muhammad. To all of them I dedicate this dissertation.

Summary

It was highlighted in a lecture by Prof. Eduardo Rojas Briales (FAO officer) at the Global Environmental and Forest Policy course of the Chair of Forest and Nature Conservation Policy, University of Göttingen, that most forest landscape negotiations exist on a very strict domestic level, barely crossing into the international utilization arena. A mere six key players steer land use regimes globally: the United States, European Union, Brazil, Russia, India and China. Meanwhile, Indonesia is listed as a secondary player. However, since the 13th session of the Conference of the Parties to the United Nations Framework Convention on Climate Change (UNFCCC), which was hosted by the Government of Indonesia and resulted in the Bali Road Map, Indonesia became an important player, initiating several important international climate change and forest landscape regime negotiations.

Being organized into five chapters, this research detects the utilization of global and regional land use regimes by national bureaucracies. This research identifies domestic political background on utilizing international regimes within specific Indonesian land use change settings, a topic which has to date been neglected in this research field, with a few exceptions (e.g. McDermott et al., 2010; Wibowo & Giessen, 2015). Investigating regimes from below became our approach to see how Indonesian domestic actors utilize land use regimes. Subsequently, the research poses several sub-questions to break down the main research question, which are as follows:

1. What are the bureaucracies and their tasks as well as their legal options to pursue actual and potential interests in steering land use transformation systems in Indonesia?
2. How did the domestic bureaucracy, together with international actors, use and adapt national instruments and international support to pursue their own (international and domestic) interests in several cases of land use transformation in Indonesia, such as (a) certification of palm oil and (b) Forest Management Units (FMU) and community forestry (CF)?
3. How did the domestic bureaucracy utilize the Association of Southeast Asian Nations' (ASEAN) regional forest and environmental regime complex?
4. How is the relevance of international and regional regimes used in domestic bureaucratic politics?

The sub-questions mentioned above were investigated and discussed using 4 constitutive articles, which can be seen in Section Publications. These publications contained nine scientific articles and four policy advisement papers

Theoretical framework

To answer our research questions, we developed a framework that uses bureaucratic politics as its main theory. This framework contributes to a description of the bureaucratic processes involved in the utilization of the selected global and regional land use regime complexes. We realized that bureaucratic politics are very strong for a bureaucratic-state-actor-based analysis, so to grasp non-actor observation, we triangulated our core theory with other theories such as power, governance, public administration, forest foreign policy and the politics of recentralisation as supplementary theories.

There are five main contents on this section. Firstly, we provide framework on explaining Indonesian dual land system on regulating conflicts and hybrid bureaucracies utilizing regimes. Secondly, we describe how's bureaucratic politics countering global private palm oil certification. Thirdly, bureaucratic politics explains the institutionalisation of forest management and community use as an instrument of state power. Fourthly, we rationalize the regional ASEAN land use regime utilities through bureaucratic politics. Fifthly, it is delivered the conceptual innovation of two political typologies such as typology on international regimes relevance and typology on domestic land use conflict visibility. Hence, we provide specific hypothesis for each case of regimes utilization that built from our theoretical framework.

Methodology

The methodological framework consists of two sections: procedures and method. The working group and multiple triangulation procedures are explained. The multiple triangulation approach was applied to support our framework for looking at international or domestic regimes from the bottom up and ensured our results were valid and reliable. The methodological framework was used for data collection and analysis of Indonesian bureaucracies utilize global and regional land use regimes. We delivered four main methods on this research such as (1) observation, (2) content analysis, (3) select interviews, and (4) case study. The author conducted three different sessions of field work such as (1) November 2013 - February 2014, (2) October 2014 - November 2014, and (3) May 2015 - June 2015) to make our non-participant observations and interviews in Jakarta, Bogor, Makassar and Jambi, Indonesia as well as in Manila, Philippines. The four methods are shortly explained below:

1. Observation: We delivered two types of observation such as non-participant observation and participant observation. We observe several many meetings, seminars, and workshops covered issues related to the Indonesian land use transformation system, which is organized by national bureaucracies and non-state actors during the field works. Non-Participant observation was also delivered by using previous personal experience of author who was involved as mediator and consultant in many practical forest policy processes and public consultations at the national, sub-national and local levels in Indonesia during period of 2005 until 2010.
2. Content analysis is applied to discover and describe the focus of individual, group, institutional or social attention in the context of actors utilizing international land use regimes. Furthermore, this method brings our analysis to be grasped as more than an individual and impressionistic interpretation. We analysed both formal and informal document (see Bowen, 2009) of the bureaucracy, which were useful in not only determining actors' positions, but also their interests.
   a. Formal document such as founding documents of institutions, legal basis, programmatic documents and other archival sources that articulated the formal responsibilities of the international institution and national bureaucracies were cited.
   b. Informal document such as internal report of bureaucracies, unpublished report, and or internal correspondence of actors were also considered.
3. Select interviews: During three sessions of fieldwork, we conducted select interviews with the relevant domestic actors utilizing international regimes and international actors who underlie regimes and national programs in Indonesia.
   a. In global and regional level, we conducted interviews with staff members of different international offices of specific organizations working in Indonesia.
   b. In national-provincial-district level, we conducted interviews with various officers of national bureaucracies as well as forestry officers from provincial and district bureaucracies
   c. We also delivered interviews with high-level staff member of private business organization such as Association of Indonesia Forest Concession Holders (APHI), National Private Association of Palm Oil Business (GAPKI), as well as non-governmental organisation's (NGO) actors.

Results

This section applies the results to our hypotheses, which were either proved or rejected based on our empirical findings; explanations are also given. Our empirical material confirms the hypotheses that national land use bureaucracies involve regional bureaucracies in order to make use of all legal options for regulating conflicts between agricultural interests and international and national conservation oriented requirements. The result also confirms hypothesis about hybrid bureaucracy steer the utilization of particular international regimes. Furthermore, it is proved that hybrid bureaucracies were created by the president were able to get financial, informational, and regulative support from international regimes. Furthermore, hybrid bureaucracies get in conflict with traditional bureaucracies and were not able to influence local land use process. This indicates that our hypotheses on formal responsibilities of land use, regional and hybrid bureaucracies have caused the fragmented and complex land use administration system in Indonesia as well as the basic bureaucratic behaviour on utilizing global and regional land use regimes.

For the Indonesian palm oil certification (ISPO) case, it is proved that in countering private global palm oil certification by establishing state-driven palm oil certification schemes, bureaucracies used their regulatory power to impose an obligation to certify. The bureaucracy's power (coercive, incentive and informational) has not only led to the successful establishment of ISPO as the new competitor, but also could be used to force voluntary schemes to accept ISPO as complementary to the global palm oil certification (RSPO) standard. This is ambiguous because some ISPO and RSPO standards conflict in the implementation of their principles, especially on legal issues. This not only enforces the reclaiming process but also successfully establishes powerful standards through legal means that affirm the global trade governance mechanism. However, our hypothesis in this case, that 'the land use bureaucracy favours the state-obligatory scheme and the trade bureaucracy favours the private-global voluntary scheme', is rejected. Our result shows different that all bureaucracies support ISPO due to the palm oil sector being transformed into a national interest. Hence, the result confirms that countering the global certification regime by creating legal barriers could be interpreted as the state hegemony using regulatory power.

Our hypotheses for the FMU-CF cases consisted of two categories, international and domestic utilities hypotheses, which are all confirmed. This shows that (1) international actors have two basic motivations for supporting FMU-CFs; formally, they want to find the most efficient way to invest their international cooperation funds in the tropical countries, and informally they want to meet their interests in maintaining the dominance of international fund providers and countering Indonesian palm oil plantations, (2) domestic bureaucracies use FMU-CF programmes to relocate power back in the central bureaucracies by preparing instruments which formal are in line with international regimes but informal are dominated by domestic bureaucracies. Finally, in a special case of recentralisation policy as a consequence of recentralising power, bureaucracies face challenges from the established decentralisation policy and its supporters in Indonesia.

The hypothesis on Indonesia utilizing ASEAN regional land use regimes is also proved, showing that Indonesia, as a powerful member of ASEAN, has four methods of utilization: (1) blocking specific rejected international initiatives, (2) attracting funding, (3) imposing standards or rules from one member state at the regional level and on other member states, and (4) allying against and/or creating a joint selling position against and/or making coalitions with other regimes. For example, in the special case of blocking utility, Indonesia Used the ASEAN Expert Groups on CITES (the Convention on International Trade in Endangered Species of Wild Fauna and Flora) and ASEAN Wildlife Enforcement Network to block the 'western-conservation' voice on listing *Ramin* timber species in Appendix I of CITES. Indonesia rejects agreement on *Ramin* timber banned due to its high economic value for Indonesia.

Finally, to contribute to international forest policy science, our hypothesis on the two-dimensional typology of the relevance of the land use regime complex is fully confirmed. The relevant typology was extensively developed and tested in our heuristic case of the Indonesian land use transformation system, using the framework of actor and problem dimensions. It is proved our hypothesis that: (1) within the type 'being relevant', although the international regimes address strong problems, national actors neglect political regulations. Informally, the bureaucracies oppose the potential solution preferred by the regimes (2) within the type 'being made relevant' national bureaucracies stress the importance of the regimes although the regimes do not address strong problems. Informally, the national bureaucracies expect to gain advantages from the regimes process.

Strategic actions for selected actors

This research will be highly relevant to the practical area of domestic utilization of global and regional regimes. Therefore, we selected and summarised three articles as strategic actions for the Indonesian delegation on ASEAN land use regimes and for indigenous actors. This research also include our direct (participant) observation on the consultation activity with *Deutsche Gesellschaft für Internationale Zusammenarbeit GmbH* (GIZ), which provides short-, medium- and long-term recommendations for Indonesia on utilizing the initiation of timber certification at the ASEAN level. This activity contains policy advice, submitted to the officials of the Ministry of Forestry and Environment, about how to use the ASEAN regional land use regime. This practical research activity also include a policy brief published by the Indonesian Communication Forum on Community Forestry (FKKM). It covers the challenges and opportunities of CF in Indonesia and provides strategic options for either strictly fighting for non-state customary forestry or alternatively supporting state customary forestry schemes.

Publications

Constitutive Publication

1. **Sahide, M. A. K.,** & Giessen, L. (2015). The fragmented land use administration in Indonesia– Analysing bureaucratic responsibilities influencing tropical rainforest transformation systems. *Land Use Policy*, *43*, 96-110. doi:10.1016/j.landusepol.2015.04.030
2. **Sahide, M. A. K.,** Nurrochmat, D. R., & Giessen, L. (2015). The regime complex for tropical rainforest transformation: Analysing the relevance of multiple global and regional land use regimes in Indonesia. *Land Use Policy*, *47*, 408-425. doi:10.1016/j.landusepol.2015.04.030
3. **Sahide, M. A. K.,** Burns, S., Wibowo, A., Nurrochmat, D. R., & Giessen, L. (2015). Towards state hegemony over agricultural certification: from voluntary private to mandatory state regimes on palm oil in Indonesia. *Jurnal Manajemen Hutan Tropika* [Journal of Tropical Forest Management], Vol. 21, (3): 162-171, December 2015. http://jesl.journal.ipb.ac.id/index.php/jmht/article/view/10798/8499
4. **Sahide, M. A. K.,** Supratman, S., Maryudi, A., & Giessen, L. (2015). Is Indonesia utilising its international partners? The driving forces behind Forest Management Units. Forest Policy and Economics. (Major revision)

Table of Content

Declaration of originality and certificate of authorship .... *viii*

Acknowledgements .... *ix*

Summary .... *x*

Publications .... *xiii*

Table of Content .... *xiii*

List of Tables .... *xiv*

List of Figures .... *xiv*

Abbreviation .... *xv*

1 Introduction and research question: International regimes from the point of view of national land use policy 1

2 Theoretical framework: Bureaucratic politics and interests in utilizing regimes 4

2.1 Formal responsibilities of domestic bureaucracies underlying interests of the land use transformation system in Indonesia 5
2.2 ISPO: From international governance to the domestic government 6
2.3 FMU and CF: The utility of national instruments for international and national interests 7
2.4 The utility of ASEAN land use regimes 8
2.5 Typology on international regime relevance 9

3 Methodology 10

3.1 Procedures 10
3.1.1 Working group procedure 10
3.1.2 Multiple triangulation procedure 10
3.2 Method 11

3.2.1 Observation ... 11
3.2.2 Content analysis ... 12
3.2.3 Select interviews ... 13

4 Results ... 14

4.1 Formal bureaucratic responsibilities underlying interests of the land use transformation system in Indonesia ... 14
4.2 The Indonesian Sustainable Palm Oil (ISPO): From international governance to the domestic government ... 16
4.3 FMU and CF: The utility of national instruments for international and national interests ... 18
4.4 The utility of the regional ASEAN land use regime complex for Indonesia ... 19
4.5 Demonstrating typologies of regime relevance used in domestic politics ... 20

5 Strategic options for selected actors ... 23

5.1 Strategic options for the MoFor delegation on ASEAN forest and land use regimes ... 23
5.1.1 Strategy for attracting funding ... 23
5.1.2 ASEAN regional timber certification: ASEAN Forest Law Enforcement, Governance and Trade Standard ... 23
5.2 Strategic options for indigenous actors ... 24

Reference ... 26

Annex: Constitutive and Additional Publications ... 33

List of Tables

Table 1 Overview of publications and contributions of each author to the constitutive publications of this study; C means core contribution and S means secondary contribution ... 3
Table 2 Multiple triangulation application ... 11
Table 3 List of activities that included as participant observation ... 12
Table 4 Documents analysed in this research ... 12
Table 5 Details of select interviews ... 13
Table 6 Main international actors engaged in FMU initiatives and their projects from 2000-present in Indonesia, and its interests (Article 4) ... 18
Table 7 Summary of domestic actors' interests in underlying FMU-CFs, which is offered to or influenced by international actors and institutions (Article 4) ... 19
Table 8 The utility of ASEAN forest and land use regimes for Indonesia (Interpreted from Article 2) ... 20
Table 9 Indicators for assessing the problem dimension of the relevance of international regimes (Article 2) ... 20
Table 10 Indicators for assessing the actor dimension of the relevance of international regimes (Article 2) ... 21

List of Figures

Figure 1 Dual system of Indonesian land administration (Article 1) ... 14
Figure 2 Legal land use transformation options in Indonesia (Article 1) ... 15
Figure 4 Relevance typology of regime complex for tropical rainforest transformation of global and regional regimes in Indonesia (Article 2) ... 22
Figure 6 Strategic options for indigenous actors considering state CFs and customary forests (adapted from Article 13 and Article 6) ... 25

Abbreviation

ADB	: Asian Development Bank
AMAN	: *Aliansi Masyarakat Adat Nusatara* or Indigenous Peoples Alliance of the Archipelago
AMS	: ASEAN member states
APHI	: Association of Indonesia Forest Concession Holders
ARKN	: ASEAN regional knowledge network
ASEAN	: Association of Southeast Asian Nations
ASEAN-FLEGT	: ASEAN Forest Law Enforcement, Governance and Trade
ASOF	: ASEAN Senior Officials on Forestry
BAPPENAS	: Ministry of National Planning
CBD	: Convention on Biological Diversity
CF	: Community Forestry
CGIAR	: Consultative Group on International Agricultural Research
CITES	: Convention on International Trade in Endangered Species of Wild Fauna and Flora
DFID	: UK Department for International Development
EU	: Europian Union
FAO	: Food and Agriculture Organization of the United Nations
FKKM	: *Forum Komunikasi Kehutanan Masyarakat* (Indonesian Communication Forum on Community Forestry)
FMU	: Forest Management Unit
GAPKI	: National Private Association of Palm Oil Business
GIZ	: *Deutsche Gesellschaft für Internationale Zusammenarbeit GmbH*
HCV	: High Conservation Value
IMF	: International Monetary Fund
ISPO	: Indonesian Sustainable Palm Oil
KPH	: *Kesatuan Pengelolaan Hutan*
MFP	: Multistakeholder Forestry Programme
MoA	: Ministry of Agriculture
MoE	: Ministry of Environment
MoFor	: Ministry of Forestry
MoEFor	: Ministry of Environment and Forestry
MoT	: Ministry of Trade
MoI	: Ministry of Industry
NGO	: non-governmental organizations
NLA	: National Land Agency
REDD	: Reducing Emissions from Deforestation and Forest Degradation
RSPO	: The Roundtable on Sustainable Palm Oil
SFM	: sustainable forest management
UNFCC	: United Nations Framework Convention on Climate Change
UNDP	: United Nations Development Programme
WB	: World Bank
WG-PATCI	: Working Group on A Pan ASEAN Timber Certification Initiative

1 Introduction and research question: International regimes from the point of view of national land use policy

Claims by international regimes regarding land use and its transformation system have been extensively analysed regarding the practices of the international actors underlying Indonesian land use transformation. The important ecological tropical forest transformative ecosystems in Indonesia have converted one million swamp areas (including peatland) for agriculture or food security (Dauvergne, 1998), timber in the 1970s (Manning, 1971), rubber in the 1990s (Feintrenie & Levang, 2009), and palm oil at the end of the 1990s (McCarthy & Zen, 2010). An international loan from the World Bank to convert one million swamps, the International Monetary Fund's requirement that the Indonesian log export ban of 1998 be lifted, the Roundtable on Sustainable Palm Oil (RSPO) governing Indonesian palm oil certification, and *Deutsche Gesellschaft für Internationale Zusammenarbeit GmbH* (GIZ) triggering the concept and implementation of FMUs and CF from the 1990s, are examples of international regimes governing the tropical rainforest transformation systems in Indonesia. We will describe the change in land use, from native forests to: i) community forestry, ii) rubber plantations and iii) palm oil plantations. This land use transformation system is our heuristic case, taken from the recent forest ecology, which was developed by the Collaborative Research Centre 990 of Georg-August University Goettingen in Germany and its counterpart in Indonesia. This heuristic model will help us to demarcate international land use regime complexes related to the heuristic case implemented and utilized by domestic actors in the policy-making process.

In scientific research, international (forest and land use) regimes are seen as the result of international relations. Researchers started by defining the regime concept (Krasner, 1983) and regime complex (Raustiala & Victor, 2004; Keohane & Victor, 2011), as well as breaking international regimes down into regional regimes (Giessen et al., unpublished; Jürging & Giessen, 2013) and quantifying regime effectiveness (which was developed by research groups at Oslo and Potsdam Universities). Forest land use in Norway was chosen by Lindsay and Solberg (2010) as a heuristic case for testing out their formula on regime effectiveness. They found it was challenging to separate the cause-effect relationship because of its complexity, including the uncertainty following longer causal chains and directions of causal influence. Moreover, Lindsay and Solberg conclude that the causal relationship of international regimes is difficult to determine and recommend analysing more complex forest management elements across countries. Furthermore, Smouts (2008) introduces 'implementation' as another criterion of effectiveness. Implementation can be measured by the way in which the provisions of the regime are translated into legislative or public policy measures.

To materialize 'implementation' and facilitate the shift from 'effectiveness' to 'influence', Bernstein and Cashore (2012) provide a framework that distinguishes four pathways influencing domestic policy making, each with its own causal logic: (i) international rules; (ii) international norms and discourse; (iii) creation of or intervention in markets; and (iv) direct infiltration of domestic policy processes. The central argument is that domestic influences cannot be studied simply by looking at the international rules pathways, as significant effects also occur along the three other pathways, as well as through interaction. This argument has pushed scholars to further analyse regimes, not only from an international viewpoint, looking into their implementation on a domestic level, but also from the viewpoint of domestic politics, utilizing international regimes and observing international actors' interests. Hence, it is necessary to balance the outlook of those regimes from national or domestic positions to asses domestic interests' utilization of those international regimes, implemented at the national level. In this framework, it is also possible to assess international actors' interests.

Regimes are studied from the bottom up in one of the new approaches developed by Working Group International Forest Regime, Chair Group of Forest and Nature Conservation Policy, University of Goettingen, which investigates the utilization of global and international regimes. Under this working group, a specific investigation of the domestic utilization of regional land use bureaucracies in the Association of Southeast Asian Nations (ASEAN) countries was delivered.

This bottom-up view requires: (i) strong domestic political analysis that investigates jurisprudential domestic land use systems that serve the potential and actual interests of related actors; (ii) investigating the selected national land use (palm oil, forest management unit and community forestry), which is strongly supported by international and domestic actors pursuing their own interests; (iii) investigating the complex utility of ASEAN land use regimes for Indonesia; and (iv) scientific measurement used in investigating domestic behaviour utilizing regimes. These four issues will use bureaucratic politics as a core theory that states that the behaviour of bureaucracies follows dual organizational interests: (1) informally they are competing with other actors, especially other bureaucracies, for resources, political domains and influence, and (2) formally, they have distinct tasks for delivering public services (Niskanen, 1971; Krott, 2005; Peters, 2010).

Hence, this research not only identifies global and ASEAN regional regimes, but also investigates the multiple actors' interests that drive domestic utilization of those regimes, which is explained on three levels. Firstly, international actors can portray their motives as rational, logical steps to disburse international funds, fight global deforestation (which means countering domestic palm oil expansion) and maintain their superiority as international donors to developing countries. Secondly, ASEAN regional land use regimes can filter global regimes that counter mutual member states' interests and/or ally with other regional regimes to attract global funds. Third, domestic actors can utilize those regimes to pursue their bureaucratic interests. Conflicting utility could also possibly exist between domestic and international actors, or conflict among them.

Following the framework of the bureaucratic politics of particular domestic programs and instruments (e.g. Indonesian palm oil certification, Forest Management Units and community forestry), which are highly supported and influenced by international regimes, further explanation will be provided regarding multiple actors, their underlying interests, and the regime concepts and their implementation. The Ministry Responsible for Forestry (MoFor) and Ministry of National Planning (BAPPENAS) have been identified as the central domestic authorities that can recentralise their power in the Ministry of Agriculture (MoA), creating a new domestic palm oil certification (ISPO) to challenge the global one (RSPO). This is in line with Wibowo et al. (2015), who stated that domestic actors utilize instruments to pursue their own interests. This research will further investigate the multiple actors to explain the regimes' utilities.

Hence, this research will contribute to shaping forestry policy by investigating the following main research question:

How do domestic politics utilize international regimes on land use transformation in Indonesia to shape forestry policy?

Consequently, the research poses the following sub-questions to break down the main research question:

1. What are the bureaucracies and their tasks, as well as their legal options for pursuing actual and potential interests in steering land use transformation systems in Indonesia?
2. How did the domestic bureaucracy, together with international actors, use and adapt national instruments and international support to pursue their own (international and domestic) interests in several cases of land use transformation in Indonesia, such as (a) certification of palm oil and (b) Forest Management Units and community forestry?
3. How did the domestic bureaucracy utilize the ASEAN regional forest and environmental regime complex?
4. How is the relevance of international and regional regimes for domestic bureaucratic politics?

The sub-questions mentioned above were investigated and discussed using 4 constitutive articles and 10 supplemental articles, as listed in Table 1. These publications contain 9 scientific articles and 4 policy advisement papers. Table 1 also shows the contributions of each author to the constitutive publications of this dissertation.

Table 1 Overview of publications and contributions of each author to the constitutive publications of this study; C means core contribution and S means secondary contribution

No.	Publications	Theory	Methodology	Results
Constitutive				
1	**Sahide, M. A. K.**, & Giessen, L. (2015). The fragmented land use administration in Indonesia–Analysing bureaucratic responsibilities influencing tropical rainforest transformation systems. *Land Use Policy*, *43*, 96-110.	S	C	C
	Giessen developed theory. Sahide and Giessen developed hypotheses of the study, methodology and methods. Sahide adapted and applied the theory and hypotheses to the case, adapted and applied the methodology to the case, and produced case findings			
2	**Sahide, M. A. K.**, Nurrochmat, D. R., & Giessen, L. (2015). The regime complex for tropical rainforest transformation: Analysing the relevance of multiple global and regional land use regimes in Indonesia. *Land Use Policy,* 47, 408-425.	S	C	C
	Giessen developed theory. Sahide and Giessen developed hypotheses of the study, methodology and methods. Sahide adapted and applied the theory and hypotheses to the case, adapted and applied the methodology to the case, and Sahide and Nurrochmat produced case findings			
3	**Sahide, M. A. K.**, Burns, S., Wibowo, A., Nurrochmat, D. R., & Giessen, L. (2015). Towards state hegemony over agricultural certification: from voluntary private to mandatory state regimes on palm oil in Indonesia. *Jurnal Manajemen Hutan Tropika* [Journal of Tropical Forest Management], Vol. 21, (3): 162-171, December 2015. http://jesl.journal.ipb.ac.id/index.php/jmht/article/view/10798/8499	S	C	C
	Giessen developed theory. Sahide and Giessen developed hypotheses of the study, methodology and methods. Sahide adapted and applied the theory and hypotheses to the case adapted and applied the methodology to the case, and Sahide, Wibowo and Burns and Nurrochmat produced case findings			
4	**Sahide, M. A. K.**, Supratman, S., Maryudi, A., & Giessen, L. (2015). Is Indonesia utilising its international partners? The driving forces behind Forest Management Units. Forest Policy and Economics. (Major revision)	S	C	C
	Giessen developed theory. Sahide and Giessen developed hypotheses of the study, methodology and methods. Sahide adapted and applied the theory and hypotheses to the case. Sahide, Supratman and Maryudi adapted and applied the methodology to the case, and produced case findings			

2 Theoretical framework: Bureaucratic politics and interests in utilizing regimes

Global and regional regimes and regime complexes

Fundamentally, an international regime can be defined as a set of implicit or explicit principles, norms, rules and procedures around which actors' expectations converge in a given area of international relations (Krasner, 1982; Giessen, 2013a). Lately, scholarly work in international relations has developed the concept of regime complexes, which are defined by Raustiala and Victor (2004:279) as 'an array of partially overlapping and non-hierarchical institutions governing a particular issue-area'. Based on this definition, the regime complex for Indonesian land use transformation can be understood as a network of institutional elements not organised hierarchically, including entire international regimes involved in a wide array of activities associated with the multiple issues involved in land use transformation (Article 2, Sahide et al., 2015).

A regime complex dedicated to Indonesian land use change might include a number of institutional elements devoted to different issues. These elements might be of a global or regional nature. In this context, regional regimes are understood as international regimes involving two or more states, with criteria limiting membership on the basis of geo-spatial or other crucial functional proximity and a lack of global aspirations (Giessen et al., unpublished; for examples, see Overdevest & Zeitlin, 2014). Hence, the concept of a regime complex affecting tropical land use transformation is proposed in this research and accounts for domestic actors utilizing those global and regional regimes to pursue their interests (Article 2, Sahide et al., 2015).

Bureaucratic politics and utilizing regimes

Utilizing regimes could be seen from two perspectives: institutionalist and rationalist. The institutionalist view argues that formal, legitimate participation in international negotiations 'could' solve some of the bureaucracies' decision-making issues (Abbott & Snidal, 1998). The rationalist perspective is more focused on the interest-based negotiation style (Müller, 2004) where elites ultimately control the political agenda (Bachrach & Baratz, 1963). We synthesized both of these approaches under bureaucratic politics, where parochial, personal, institutional, national and domestic interests influence the actors' positions on the issue at hand during participation in international negotiations (Tagma & Uzun, 2012).

We used bureaucratic politics as our core theory to investigate actors' behaviour and find their basic interests in the particular utilization of international regimes. This is find in Article 1, 2, 3, and 4 or Sahide and Giessen, 2015; Sahide et al., 2015a; Sahide et al., 2015b, Sahide et al., 2016). Bureaucratic politics have two major assumptions: (i) states are not fully rational and unitary actors and (ii) the central state lacks the capacity and information to assess and control the increasing amount of international interaction (Niskanen, 1971; Peters, 2010; Krott, 1990; 2005; Rahman and Giessen, 2014; Burns and Giessen, 2014). Most of the players utilize foreign policy decision-making by virtue of their roles and positions in the government (Allison, 1971). Therefore, to understand roles and positions, interests should be at the centre of analysis, complementing the formal tasks of bureaucracies (Krott, 2005). Brockhaus et al. (2014) strengthen this framework by arguing that by obtaining, formally and informally, the consent of the governed, actors that are defined as individuals, groups of people, or organisations with the capacity and legitimacy to exercise their bureaucratic power, do so when political interests are present. In addition, to elucidate these interests, we look into power, particularly coercion power, arising from bureaucratic politic entities' tendency to use their influence (which is exerted through the use of information and power) to steer developments in their desired direction (Simon, 1981; Krott, 1990; 2005).

Bureaucratic political theory involves not only state actors, but non-state ones as well (global and domestic private, third-party or indigenous actors), who are defined as bureaucracies' societal clientele. There are also arenas where bureaucracies struggle to increase their responsibility for policy domains, budgets and staff, as well as maintain mutually beneficial relations with their societal clientele (Downs, 1967; Peters, 2010; Giessen et al., 2014).

In hindsight, we used bureaucratic politics to observe the multiple utilities available to domestic actors using the international regime in land use politics. Consequently, this research will create a new arena of scholarship, 'regimes examined from below' or bridging regimes and domestic politics to fill the gap in materializing regimes and their implementation in specific national settings', which has been neglected in research, with a few exceptions (e.g. McDermott et al., 2010). We will analyse multiple actors' informal interests, which drive their formal performance. In a specific context, underlying domestic interests could lead to imposing a political agenda, aligning with other actors, blocking annoying agreements, or taking other bureaucratic politic actions involved with utilizing regimes.

2.1 Formal responsibilities of domestic bureaucracies underlying interests of the land use transformation system in Indonesia

Explaining dual land system on regulating conflicts through bureaucratic politics

Indonesian state forests were influenced by colonial-era state making (Peluso & Vandergeest, 2001) both in terms of territorialisation of the dual land system of state forests and land titles (Fay & Martua, 2005) and as a legal framing for separating forest governance into beheer (administration) and bestuur (management) (Article 1). The two colonial forest practices, the dual land system and the separation of forest governance, have directly impinged on the intended use of forests, even as they were defining what these were under the law (Peluso & Vandergeest, 2001). Under the bureaucratic politic framework, the dual land system and forest governance separation have put land use bureaucracies (e.g. national land and forestry ministries) in place that have improved and strengthened the formal legal land- and commodity-based strategy controls of the state. This situation should be understood as bureaucracy steering legal land usage to pursue their interests.

Under the old forestry and agrarian laws (Basic Forestry Law 5 of 1967 and Basic Agrarian Law 5 of 1990), the national land use bureaucracies has fully administering forest and land use system, where Ministry of Forestry (MoFor) held practically full authority to administer an area of 143 million hectares (ha) that had been determined as 'Forest Area' (*Kawasan Hutan*) and National Land Agency (NLA) administering land territoriality or 'Non-Forest Area'. However, under new forestry (Forestry Law 1999) and decentralisation laws, national land use bureaucracies has to involve regional bureaucracies in order to make use of all legal options for regulating conflicts between economic and conservation interests (see Barr, 2006). The MoFor will use forests determined by law to counter international forest regimes in defining forest and land degradation. In addition, the forestry and land bureaucracies are engaged in a tenacious struggle regarding authority over the land (Nurrochmat et al., 2014), but could be on the same side when disputing local land conflicts (Bakker & Moniaga, 2010). The competition between land use bureaucracies and agricultural commodity bureaucracies is intensified by the interests of the societal clientele surrounding each bureaucracy (see Peters, 2010; Downs, 1967). Land use bureaucracies are involved with and influenced by societal clientele through their political actions, such as imposing power during the authoritarian era, involving local (provincial and district) bureaucracies during the decentralisation period, and recentralising power during the post-decentralisation era.

Hybrid bureaucracies utilize particular international regimes

Public bureaucracies provide an additional point of access for a wide variety of interests in a fragmented government (Thomas, 1997). The fragmented interests of bureaucracies can be naturally divided into production-oriented bureaucracies and conservation-oriented bureaucracies (Hirsch & Warren, 1998). It is possible, however, to find hybrid production-environmental bureaucracies (Giessen et al., 2014). According to public management scholars, the bureaucratic hybrid structure is the best form for organizing and influencing multiple clients' interests (Pedersen, 2010). For our framework, a hybrid bureaucracy is considered to convey different interests in the global and domestic agendas. The medium status of the legal basis, but strong power, of hybrid bureaucracies will meet the high-level legal challenges of traditional bureaucracies (almost all of which are clearly divided into production- or conservation-oriented bureaucracies).

These strong hybrid bureaucracies can manifest as a super bureaucracy, but on a temporary, state level where organisations are established by the elected president to pursue the temporary and relative interests of the national agenda. Creating a new super-bureaucracy with limited staff but strong coercion power and a budget is easy (Helperin & Clapp, 2006; Wibowo and Giessen, 2015), but it is not an effective way to face traditional bureaucracies with their long-established, clear forest and land use territoriality. Hence, contested interests and power struggles between bureaucracies provide opportunities to utilize international regimes for pursuing domestic interests.

Hypotheses

- Hypothesis I: National land use bureaucracies involve regional bureaucracies in order to make use of all legal options for regulating conflicts between agricultural interests and international and national conservation oriented requirements.
- Hypothesis IIa: Hybrid bureaucracies were created by the president to steer the utilization of particular international regimes.
  Hypothesis IIb: Hybrid bureaucracies were able to get financial, informational, and regulative support from international regimes.
  Hypothesis IIc: Hybrid bureaucracies get in conflict with traditional bureaucracies and were not able to influence local land use process

2.2 ISPO: From international governance to the domestic government

Certification as voluntary private governance?

In the 1990s, certification in the agrarian policy domain was introduced (Rametsteiner & Simula, 2003). Studies conceptualized certification as 'non-state market-driven governance' and a 'governance system without government' (Cashore, 2002). Cashore et al. (2007) argue that 'the most important feature of non-state market-driven governance is that there is no use of state sovereignty to enforce compliance'. However, recent scholars note that under global circuits, the state has reassumed the government's role in adapting public services and new institutions emerged for different motives, likely following political change (Cordoba & Jansen, 2014). Furthermore, Hospes (2014) found that the difficulties in the implementation of the voluntary private standard at the domestic level (in the southeast Asia) have prompted the collaboration of scholars with national or local governments. To synthesize this logical framework, we pose the following question: Is non-state market-driven governance purely private or state driven? (Article 3)

Bureaucratic politics in countering global certification

Certification politics among inter-specific state bureaucracies and governing conflicts among non-state and state actors can be further specified into which particular bureaucracies and their societal clienteles are conflicting with other bureaucracies and their non-state allies (see Peters 2010). Bureaucratic politics theory claims that state bureaucracies compete with each other for resources, staff and responsibility for policy domains (Krott, 2005; Peters, 2010; Sahide et al. 2015a; Giessen et al., 2014, Burns et al., 2016). In order to increase their power, national bureaucracies are likely to seek coalitions with national and international actors, both public and private (Rayner et al., 2001; Giessen & Krott, 2009). Recent findings show that mandatory state certification was introduced through legal state verification schemes in the timber case, where national bureaucracies could expand their authority through national sovereignty (Schouten & Glasbergen, 2011; Cashore & Stone, 2012; McDermott, 2013).

The Indonesian palm oil sector competes with the agricultural-trade bureaucracies and land use bureaucracies (Sahide & Giessen, 2015), which is important to investigating their preferable coalition for private or state certification. Giessen et al. (2014) show that utilitarian bureaucracies (e.g. trade and industrial ministries) are strongly represented in state interests rather than specialized environmental bureaucracies in international forest and land use regimes. Following this logic, land use bureaucracies could favour a state palm oil certification and trade and industrial ministries could prefer global regimes and represent the domestic level in international negotiations. In this sense, land use bureaucracies could enhance certification schemes by funding program participation, promoting certification through public procurement policies or applying certification standards to their own practices (Espach, 2006).

To use a different logic, it is should be considered that bureaucracies, in this case as the monopolists holding regulatory power (Wilson, 1975), are explained by our bureaucratic politics model as rational-legal bureaucracies following national and institutional interests. Particularly, when bureaucracies become less fragmented (Boyne & Cole, 1998) and meet mutual interests or even become national interests, they will act as the sole, solid bureaucratic provider of a service (Osborne, 1993). Countering the global certification regime by creating legal barriers could be interpreted as the state hegemony using regulatory power (see Lazarus-Black, 1994) in such political actions as imposing rules to tackle global schemes, lowering the domestic standards of certification institutions, making certification mandatory, and imposing cooperation through joint technical assessment.

Hypotheses

- Hypothesis III: In countering private global palm oil certification by establishing state-driven palm oil certification schemes, bureaucracies use their regulatory power to impose an obligation to certify and try to attract global third-party actors to acknowledge national certification.
- Hypothesis IV: Different bureaucracies favour different certification schemes based on their bureaucratic interests. Land use bureaucracy favours a state-obligatory scheme and trade bureaucracy favours a private-voluntary scheme.

2.3 FMU and CF: The utility of national instruments for international and national interests

Bureaucratic politics explains the institutionalisation of forest management and community use as an instrument of state power

The post-colonial era led to hollow states in Indonesia when the state failed to govern forest management due to its unclear tenure (Ostrom, 1999; Agrawal, 2007). Therefore, the institutionalisation politics of forest management have been used as a technic of state power (Agrawal, 2001 also gives an example from India) to not only overcome this issue, but also to meet specific domestic interests. Institutionalising forest management is part of the territorial control process (Peluso & Vandergeest, 2001) by which governments demarcate specific functions of forest land (e.g. conservation or production) and also mark areas of forest land as claimed by the state or potentially offered to private and indigenous actors but still under state control. Non-state indigenous actors might be against forest management institutionalisation if they wish to promote an indigenous concept that is purely established and free of any state structure (Bakker & Moniaga, 2010). This is also in line with Peluso and Vandergeest (2001), to whom access to state forests and forest resources was provisionally granted through state-issued permits that considered their research as coming under CF usage.

Rich technocratic instruments of forest management institutionalisation in a 'hollow forest' management situation are easily developed and clearly linked to international forestry norms. Gourevitch (2009) shows that international norms can became an explanatory variable for domestic politics. With the support of technocratic societal cliental through consultancy programmes (Hamilton-Hart, 2006), states can easily adapt these norms into very normative and legal tools, involving other international actors supporting institutionalised forest management. Rich technocratic forest management institutionalisation is interpreted by bureaucratic politics (Krott, 2005) as a potential strong state power in the new bureaucratic structures, with budget essential-transferred and mechanisms, capacity, and informational requirements and management. Consequently, diverse domestic utilization of this forest institution can be used to relocate power back in the central bureaucracies, like recentralising authority (Mutebi, 2004; Ribot et al., 2006) or recentralising budgets and information.

Following bureaucratic politics to explain societal clientele (Downs, 1967; Peters 2010), this should be understood as the bureaucratic opportunity for actors in the forestry and land use sectors to maintain some balance between state rule, policy domain and coercive pressure from indigenous actors regarding customary forests rights. Indigenous actors, who are always strongly allied with the international conservation regime (Anaya, 2004), could reject the state's proposal on community forestry; however, the international forest regime's majority will also utilize this domestic forest institutionalisation to infiltrate domestic policy making through direct access (Berstein & Cashore, 2012). They will then pursue their formalistic regime's goals and informal interests; and vice versa, domestic actors will utilize this instrument as an expressive instrument for preparing direct access to inviting regimes (similar to McDermott et al., 2010).

Hypotheses

- Hypothesis V: International actors have dual motivations for supporting FMU-CFs: (i) formally they want to find the clearest and efficient way to invest their international cooperation funds in tropical countries and counter global deforestation (ii) informally they want to counter Indonesian palm oil plantations. In addition, international interests could be contrary to domestic interest in utilizing FMU-CFs
- Hypothesis VI: Domestic bureaucracies use FMU-CF programmes to relocate power back in the central bureaucracies by preparing instruments which formal are in line with international regimes but informal are dominated by domestic bureaucracies.

2.4 The utility of ASEAN land use regimes

ASEAN regional regimes

Different from European Union (EU)-style regional integration, the Association of Southeast Asian Nations (ASEAN) has tended to be more and less intergovernmental rather than supranational in authority (Cuyvers et al., 2005). Furthermore, the nature of the actors' relationships in achieving member goals in the ASEAN regime has been identified as a very soft diplomacy or commonly cited as 'the ASEAN way' approach to maintain its traditional political doctrine on striving for peace and security in Southeast Asia (Acharya, 1998; Lian & Robinson, 2002; Cuyvers et al., 2005), or one could say it is even less conflict-inducing than the non-intervention principle (Goh, 2003).

Explaining ASEAN land use regime utilities through bureaucratic politics

A non-intervention code is a strong instrument for making 'process' (not 'progress') in the ASEAN regimes (Jones & Smith, 2007). It characterises regional land use regimes and bureaucratic political analysis explains that strong domestic sovereignty is the convincing instrument and resource of state power that allows domestic bureaucrats to pursue their interests by blocking any irrelevant regional treaties (see Krott, 2005; Peters, 2010). Rather than creating an integrated forest and environmental community (Cotton, 1999), it can only sustain a pattern of limited intergovernmental relation. Domestic actors can easily be bureaucratically rigid and normative collaborators in their relations with other actor states' members, which in international politics is defined as 'political delay' (Huntington, 1965). It occurs when one bureaucracy lacks autonomy and coherence. It is intensified by the contrasting interests of different domestic bureaucracies involved in one region's forest negotiations; for example, agricultural ministries versus forest ministries at the ASEAN Ministerial Meeting on Agriculture and Forestry (AMAF) forum. Conflicts of interest between agricultural, environmental, trade and forestry bureaucracies will enhance the actors' representation in the regional regime so they only use regimes as a normative end to achieve their bureaucratic interests, or to maximize budgets, expand staff, collect information and observe opposing interests (see Niskanen, 1971; Krott, 2005; Peters, 2010) without deeply negotiating on behalf of either production or conservation interests. However, domestic conservation interest bureau representatives will still lose when production interests meet over the mutual benefits of state members.

In hindsight, this ASEAN regional land use regime study has resulted in some researchers observing the ways of utilities for ASEAN's members using the regional regime in forest and land use politics. Firstly, it is observed that ASEAN has undermined forest and environmental issues (Elliott, 2000; Aggarwal & Vakkey, 2013), which means blocking inconvenient forest and environmental issues to benefit specific member(s) interests. Secondly, specific ASEAN member(s) or other regimes could potentially impose issues on other member state(s) through ASEAN bureaucracy; it is similar to the Yong & Peh (2014) thesis that ASEAN, in tackling forestry and environmental issues, cannot be underestimated, which can be seen in the example that all ASEAN members have finally ratified and are enforcing the Agreement on Transboundary Haze Pollution, which is yet to be accomplished in any other regional regime in the world. Thirdly, a significant amount of the budget is produced from the international actors' support for ASEAN, which means ASEAN is attracting funds through their forest and environment issues. Fourthly, ASEAN member states (AMS) need a strong alliance to face other regimes in order to maintain their urgent position on global forest and environmental politics.

This starting point provides the framework to explain actors' behaviour for producing the comprehensive implementation for analysing AMS' behaviour using ASEAN for blocking, attracting, imposing, and allying to maintain their formal and informal goals.

Hypothesis

- Hypothesis VII: A regional regime is useful for a powerful member as given in the case of Indonesia in four different ways: (1) blocking specific rejected international initiatives; (2) attracting funding; (3) imposing standards or rules from one member state at the regional level and/or on other member states; and (4) allying with and/or taking a joint selling position against and/or forming coalitions with other regimes.

2.5 Typology on international regime relevance

A typology in this field poses scholars of reasonable politics with a number of conceptual challenges. These call for conceptual innovation in the area of comparative land use regime complex analysis (Wigell, 2008; Guliyev, 2011). Regarding international politics, regime analysts need to solve the problem of creating analytic differentiation of diverse forms of political land use regimes without stretching their concepts to cases that do not display reasonable conceptual validity (Wigell, 2008). Scholars have stressed the importance of re-thinking key concepts found in the literature on political land use regimes and domestic land use politics in order to allow for typologies that can better describe these new political applications as a better basis for cross-national land use comparison practices.

Configurationally, regime typologies became the major tools that could be generalized across the wide variety of international regimes (similar to Ebbinghaus, 2012 on welfare state regimes). However, as such they unavoidably share some disadvantages, so that the limitations outweigh the contributions of some typologies (see Elman, 2005; Wigell, 2008) depending on the particulars of their formulation and the ways in which they are employed when both a tropical land use regime complex exists. Our typology is an analytic idea intended, in the Weberian bureaucratic focus tradition (Elman, 2015; Guliyev, 2011), as a typology of regime relevance in the study of domestic politics. Establishing the relevance of a land use regime, which is large and complex, needs facilitation of typological reasoning and compression. Elman (2005) argues that these facilitations, working with multivariable explanatory typologies, would allow the analyst to discover missed combinations and suppressed assumptions of regime relevance. We propose a typology of regime relevance in the new study of bridging international and domestic politics.

Fundamentally, we propose two explanatory dimensions of typology categories, actor and problem dimensions, by which a two-dimensional typology is better grounded for mapping regimes, particularly hybrid and complex regimes (Wigell, 2008). Four pathways or regimes—here we were influenced by Bernstein and Cashore (2012)—implicitly suggest that international regimes are actually relevant to their member states. This ignores the option of treating such regimes as merely symbolic policy. Hence, what these novel approaches underestimate is that in any given domestic context a regime may or may not be relevant in two ways. This two-dimensional typology tends to (i) understate and review the coherence problem between a regime's scope and domestic land use problems and (ii) analyse actors' behavioural actions using bureaucratic politics for utilizing the global and regional land use regime complex. In this two-dimensional approach, four types of relevance are postulated: 'pure relevance', 'made relevant', 'being made relevant' and 'non-relevant'. Furthermore, this two-dimensional approach tends to minimize the complexity of the land use regime complex. However, for future research, the dichotomous dimensions approach should be investigated to see if there exists an overlap between different types of regime (see Wigell, 2008).

Conclusively, in anchoring our typology, we used a simple quantitative approach (see Wallace, & Singer, 1970) for measuring the actor and problem dimensions using a bureaucratic politics approach. Firstly, the problem addressed by a particular regime may be pertinent specifically to a given country, leading to the relevance of that regime in a **problem dimension** (Krasner, 1982; 1983). Indeed, multiple justifications are a source of broad attraction in regime scope, whether it is correspondent or forced to be correspondent with domestic priority issues. Secondly, regimes may be made relevant in a given domestic context due to their active use by domestic policy actors in order to justify their preferences, even if the problem the regime is addressing does not pertain to the country in question. The latter situation is rooted in interest-based bureaucratic politics theory (Krott, 2005; Peters, 2010), according to which nation-level bureaucracies may utilize international negotiations, regimes and actors to further their interests domestically (Giessen et al., 2014; Giessen, 2013; Arts, 2012; Arts et al., 2013). Here, this is considered to be the **actor dimension** of regime relevance. In the motivations of key bureaucratic actors, we are focusing our attention on how they maximize budgets and staff to support regimes, support of a high legal basis, and other bureaucratic resources. There is much spillage between regime opportunities that are available and domestic actions that are preferable, and we explore how these regimes prefer to be utilized by domestic actors or regimes, being either fully implemented or blocked/partly blocked, which will explain their basic interests.

Hypotheses

- Hypothesis VIII: Using the two dimensions: (1) actor and (2) problem, a typology of regime-relevance discriminates four types of relevance: 'pure relevance', 'being relevant', 'being made relevant' and 'non relevant'.
- Hypothesis IXa: Within the type 'being relevant', although the international regimes address strong problems, national actors neglect political regulations. Informally, the bureaucracies oppose the potential solution preferred by the regimes.
  Hypothesis IXb: Within the type 'being made relevant' national bureaucracies stress the importance of the regimes although the regimes do not address strong problems. Informally, the national bureaucracies expect to gain advantages from the regimes process.

3 Methodology

3.1 Procedures

3.1.1 Working group procedure

This research was a part of Working Group International Forest Regime, Chair Group of Forest and Nature Conservation Policy, University of Göttingen. Therefore, the working group discussion produced a general methodology and hypotheses based on the core theories applied, including bureaucratic politics and international regimes. Afterward, each individual researcher developed their detailed methodology and hypotheses and triangulated them with other theories based on specific cases of Indonesia's utilization of the land use regime complex. Specific literature from other colleagues in the working group was used (e.g. Wibowo & Giessen, 2012) as a starting point. This is in line with Sadath and Krott's (2012) assertion that issues can be a starting point in policy analysis.

3.1.2 Multiple triangulation procedure

Triangulation refers to the use of more than one approach to investigate a research question in order to increase the validity of the research results (Denzin, 1970; Wilson, 2014). Data, investigator, theory and method triangulation supported our framework for looking at international or domestic regimes from the bottom up and ensured our results were valid and reliable. The multiple triangulation approach was utilized to reduce author bias and subjectivity.

As shown in Table 2, the establishment of the new government of Indonesia in October 2014 has potentially established a new configuration of bureaucratic politics regarding Indonesia's land use regime change system. Hence, three field works were conducted concerning the timeframe of domestic political issues such as before the new government was established (December 2013 until February 2014), during the new government's establishment (October 2014 and November 2014), and after new government's establishment (May 2015 and June 2015) to cover the data changes and to increase the reliability. To comply with different level of actors and bureaucracies, it is collected data from national level in Jakarta and Bogor, sub national level in South Sulawesi and Jambi provinces, as well as in ASEAN level in Jakarta and Manila-Philippines.

Table 2 Multiple triangulation application (Article , Article 2, Article 3 and Article 4)

Multiple triangulation		Detailed
1. Data triangulation	Three different time of field works (9 months), reflecting different socio-political situations	Non-participant observation: 2013-2015 1) Before presidential election (the previous central government) (November 2013 until February 2014) 2) During the new government establishment (October 2014 and November 2014) 3) After presidential election and establishment of new cabinet (May 2015 and June 2015)
		Participant observation: 2003-2012
	Different locations	1) National level in Indonesia 2) Sub National level (province and district level) in Java island and outside Java island 3) Outside Indonesia in the Philippines
	Different respondents	1) State bureaucracies 2) Non-state actors 3) Inter-state (regional) bureaucracies
2. Investigator triangulation	Use of more than one researcher	Using local expert university researchers in Indonesia and a forest researcher in United States
3. Theoretical triangulation	Using core theories, and supplementary theories	1) Bureaucratic politics as the core theories. 2) International regimes, power, governance, public administration, forest foreign policy, and the politics of recentralisation as the supplementary theories
4. Methodological triangulation	Use of more than one method	1) Observations, 2) select interviews, 3) content analysis

To ensure the reliability of our data interpretation, we involve local expert universities and a forest researcher in United States as the investigator triangulation. In addition, we triangulate bureaucratic politics theory with other theories to deal with specific investigation of particular regimes utilization. Furthermore, to ensure our analytical results were valid, it is used three different methods such as observations, select interviews, and content analysis.

3.2 Method

3.2.1 Observation

Non-participant observation (Article 1, 2, 3 and 4) is a very strong method in political analysis (Verd, 2004; Fenno, 1986; Patton, 2005; Van Evera, 1997) and was particularly apt given our focus on actors' interest (Verd, 2004) in utilizing regimes. The author conducted three different sessions of field work (see Table 2 for the frame time triangulation) to make our observations and record opinions, interests and alliances made by the actors utilizing global and ASEAN regional regimes in several national and international forest and land use meetings in Jakarta, Bogor, Makassar and Jambi in Indonesia and Manila, Philippines. The meetings, seminars, and workshops covered several issues, such as land grabbing (1-2 December 2014), public consultation from ADB for the FMU project (11 June 2015), community forestry and customary forestry reflection (27 November 2013 and 8 December 2014), ideas for the National Development Plan 2015-2019 (3 December 2014), Indonesian timber certification (15 January 2014), practical palm oil development in Indonesia (9 October 2014), the ASEAN community (1 October 2015), synergizing timber certification with REDD+ (22-23 December 2011), the outlook and reflection of the forestry sector (5 December 2013), and several meetings of non-governmental organizations (NGOs) concerning forest resource governance. We observed the informal manoeuvring behind actors' behaviour during and before the events, and we followed up with informal meetings with selected actors after the events to observe particular informal interests of actors.

Previously, the authors were involved as mediator and consultant in many practical forest policy processes and public consultations at the national, sub-national and local levels in Indonesia. During period of 2005 until 2010, the authors mediate conflict on Bantimurung Bulusaraung national park including represented as facilitator on legal drafting process of community forestry regulation in Maros District in 2005. In addition, the authors also acted as consultant on developing several FMUs and CFs in Bantaeng and Maros districts in 2009 until 2010. The authors also acted as sub-consultant for GIZ on stock taking study for ASEAN timber legality in 2015. We used these personal experiences as the participant observations (Article 4). Participant observations means the researcher have their own interest representing their role such facilitator and conflict mediator (see Atkinson &

Hammersley, 1994). These observations (accompanied by interviews) helped identify the actors' interests and were key in collecting and verifying empirical findings (Van Evera, 1997). Detailed participant observation can be seen in Table 3.

Table 3 List of activities that included as participant observation (Article 4)

Activity	Period	Actor involved	Role
Legal drafting of Maros District Regulation on Community Forestry	October – December 2005	Forestry Agency of Maros District, Maros Parlement, NGOs, Hasanuddin University	Facilitator
Conflict resolution on Bantimurung Bulusaraung National Park	2010-2011	Forestry Service of Maros District, Maros Parlement, NGOs, Hasanuddin University	Mediator
Establishing Village Forests in Bantaeng District	2009-2010	Labbo and Patenetang village leaders, Forestry Agency of Bantaeng District, NGOs, Hasanuddin University	Facilitator and Consultant
Public Consultation on Establishing Forest Management Units in Bantaeng District and Maros District	2009-2010	MoFor representatives, Provincial NGOs, Forestry Agency of South Sulawesi Province, Forestry Agency of Bantaeng District, Hasanuddin University	Facilitator and Consultant
Regional Facilitator for South Sulawesi Province, Indonesian Communication Forum on Community Forestry (FKKM)	2006-2009	National and Provincial NGOs, Forestry Agency of South Sulawesi Province, Hasanuddin University	Facilitator
Stock Taking Study on Existing Initiatives for Ensuring Legal and Sustainable Timber in ASEAN	March – July 2015	GIZ, the delegation from ASEAN Member States, ASEAN Secretariat	Consultant

3.2.2 Content analysis

Content analysis (Article 1, 2, 3, and 4) is a commonly used qualitative research technique (Hsieh & Shannon, 2005). This method allows the researcher to examine large quantities of data with relative ease in a systematic manner (see Stemler, 2001). Therefore, content analysis enabled us to discover and describe the focus of individual, group, institutional or social attention (see Weber, 1990) in the context of actors utilizing international land use regimes. Furthermore, this method brings our analysis to be grasped as more than an individual and impressionistic interpretation (Anderson et al., 2006). To identify actors and bureaucracies as well as their formal responsibilities and their positions, we analysed data from diverse sources, such as policy documents, media reports and professional journals published by political actors, which were the basic resources for qualitative content analysis, a flexible method for analysing textual data (Hsieh & Shannon, 2005; Neuman, 2005). Furthermore, we analysed both formal and informal document (see Bowen, 2009) of the bureaucracy, which were useful in not only determining actors' positions, but also their interests.

Formal document such as founding documents of institutions, programmatic documents and other archival sources that articulated the formal responsibilities of the international institution and national bureaucracies were cited. We analysed formal documents that were published by forest and land use related international organization (e.g. UNFCC, UNDP, RSPO, FAO, CITES, Convention on Biological Diversity/CBD) during period of 2010-2015. National forest and land use legal basis (e.g. laws, presidential regulation, ministerial degrees, provincial and district regulations) since 1960s until 2015 were also collected. We collected programmatic reports, annual and performance report, leaflet and budget flows from various national bureaucracies (e.g. MoFor, MoA, NLA), which is published in 2010-2015. We also considered consultancy reports from GIZ, ADB, and RECOFTC for FMUs and CFs development that were published in 2012-2015. We also analysed informal report on MoFor relating to the actor's interests in utilizing ASEAN land use regimes, which is provided by our respondent. Additional information regarding these documents can be seen in Table 4.

Table 4 Documents analysed in this research (Article 1, 2, 3, and 4)

Documents	Period	Publisher
Report on some international organization activities (UNDP, UNCTAD, UNFCC, UNDP, UNESCO, FAO, CITES, Wetland International, RAMSAR, CBD, FSC, PEFC	2010-2015	UNCTAD, UNFCC, UNDP, UNESCO, Wetland International, RAMSAR, FSC, PEFC, CITES, CBD, FAO
RSPO news, leaflet, budget and activities	2010-2015	RSPO
ISPO news, leaflet, budget and activities	2010-2015	MoA, ISPO Commission,
Internal report for ASEAN regime utilization	2012	
Report and activities of some working group on forest and environmental of ASEAN bureaucracies	2012	ASEAN Secretariat

International forest cooperation and activities in MoFor	2012-2013	MoFor
Annual reports and performance reports of related ministries (MoFor, Ministry of Home Affair, MoA, Ministry of Trade, Ministry of Industry)	2012-104	MoFor, Ministry of Home Affair, MoA, Ministry of Trade, Ministry of Industry
Constitutional court decision, Presidential regulation, Ministerial regulation, ministerial decision, provincial and district regulation,	1960-2015	Constitutional Court, MoFor, Ministry of Home Affair, MoA, Ministry of Trade, Ministry of Industry
Reports on FMU development from Asian Development Bank and GIZ	2012-2015	GIZ and Asian Development Bank
Reports on FMU and CF development from RECOFTC and Hasanuddin University	2004-2015	RECOFTC and Hasanuddin University

3.2.3 Select interviews

We conducted select interviews (Article 1, 2, 3, and 4) with the relevant domestic actors utilizing international regimes and international actors who underlie regimes and national programs in Indonesia. Domestic actors came from forest bureaucracies, agricultural bureaucracies, trade and industrial bureaucracies, super bureaucracies, provincial-district bureaucracies, the private sector, NGOs, and indigenous communities. International actors were officers or consultants of specific foreign international corporations working in Indonesia in connection to land use programs, and staff from ASEAN working groups or forums. The purpose of these select interviews was to develop a causal explanation for the hypotheses that resulted from a framework or hypothesis testing (Mosley, 2013).The different interest-based bureaucracies and actors interviewed increased the validity of our data on domestic utilizing regimes. Different opinions, arguments and positions increased our understanding of specific actors' interests through crosschecking with their opponent actors or competitors from one specific regime or program implemented in Indonesia.

In global and regional level, we conducted interviews with staff members of different international offices of specific organizations working in Indonesia (e.g. GIZ, MFP/DFID), consultants of the ADB's project during the third fieldwork (2015), and staff members of ASEAN secretariat during the second and the third field work (2014 and 2015). In national-provincial-district level, we conducted interviews with various officers of national bureaucracies (e.g. MoFor, MoA, NLA, Task force for REDD+, Ministry of Trade, Ministry of Industry) as well as forestry officers from provincial and district bureaucracies during the second field work in 2014. We also delivered interviews with high-level staff member of private business organization such as Association of Indonesia Forest Concession Holders (APHI) and National Private Association of Palm Oil Business (GAPKI), as well as NGOs actors such as AMAN and FKKM during the first and the second field work in 2013 and 2014. Details on the actors interviewed and the interview time frame are provided in Table 5.

Table 5 Details of select interviews (Article 1, 2, 3, and 4)

Interviewee and location	**Time frame**	**Topic: Type of actors utilizing specific regimes**
I. Global and regional actors		
One staff member from *Deutsche Gesellschaft für Internationale Zusammenarbeit GmbH* (GIZ), one staff member of Multistakeholder Forestry Pro gramme (MFP), staff from the MFP 3 (DFID), two consultants of the ADB's project for the multifarious FMU-CF program and ASEAN projects in Indonesia	01.06.2015, 02.06.2015, 10.06.2015, 18.06.2015	Global actors underlying FMU-CF and ASEAN regime utilization
One officer from the RSPO Indonesia Office and one officer from Greenpalm (via email). Germany and Indonesia	05.02.2014, 07.06.1013, 18.07.2014	Global actors utilizing the palm oil certification regime
One officer from Malaysia, one officer from the Philippines, representatives at ASEAN meetings in Manila, Philippines	04.06.2015	Regional actors utilizing ASEAN regimes
II. National-provincial-district bureaucracies		
One official from the Presidential Task Force Unit in Jakarta and, via email, in Germany, and one officer from the National Forestry Council. Indonesia	11.02.2014, 20.02.2014, 07.02.2014, 19.01.2014	Super bureaucracy utilizing the global and ASEAN regional land use regimes (in general)
Four general officers from the Ministry Responsible for Forestry (MoFor), two officers representing the MoFor at the ASEAN forest and environmental forum, and two expert staff members from the MoFor. Indonesia,	19.12.2013, 15.01.2014, 12.02.2014, 11.02.2014	Forest and land use bureaucracy utilizing regimes such as the ASEAN regime, indigenous regime, research regime, climate regime, and timber and rubber regime
One officer from the ISPO Commission and one officer from the MoA. Indonesia.	23.01.2014, 24.01.2014	Agricultural bureaucracy utilizing palm oil, rubber and agricultural trade regimes

One officer from the Ministry of Trade who was in charge of the national timber mandatory certification. Jakarta, Indonesia,	12.02.2014, 11.02.2014	Trade and industrial bureaucracies utilizing illegal logging, and timber trade regimes
One high-level staff member of provincial bureaucracies, via e-mail, and two from district bureaucracies utilizing FMU-CF in Nusa Tenggara Barat and South Sulawesi Provinces, via e-mail in Germany	26.04.2015, 30.04.2015	Provincial and district bureaucracies utilizing regimes and FMU-CF
III. Private and NGO actors		
One high-level staff member of the Association of Indonesia Forest Concession Holders (APHI), National Private Association of Palm Oil Business. Jakarta and Bogor, Indonesia.	19.12.2013, 20.01.2014	Private actor utilizing forest, rubber and palm oil regimes
Two staff members of different NGOs who were affiliated with global certification, and NGOs implementing one of the ITTO projects in Indonesia. Bogor, Indonesia	22.01.2014, 17.01.2014	Non-state actors utilizing timber certification regimes
One leader of the Indigenous Peoples Alliance of the Archipelago (AMAN) in Bogor Indonesia, and via e-mail, one staff member of an NGO who is facilitating state and non-state CF in South Sulawesi province via e-mail in Germany	11.02.2014, 23.04.2015, 15.05.2015	Domestic indigenous actors utilizing international regimes and countering FMU-CF

4 Results

4.1 Formal bureaucratic responsibilities underlying interests of the land use transformation system in Indonesia

One part of hypothesis I is confirmed: land use bureaucracies, such as the MoFor and the National Land Agency (NLA), involving regional governments are the core bureaucracies responsible for strengthening the dual land system. The MoFor acts as the main administrator for state forest areas and due to the provincial- and district-based autonomy the regional government also has a fragmented role in the land use administration system (see Figures 1 and 2). This finding also enhances the findings of Bakker and Moniaga (2010) and Fay and Martua (2005) that Indonesia has a complex legal system that is operationalized by the strong dual land system of forest state areas and land titles.

Figure 1 Dual system of Indonesian land administration (Article 1)

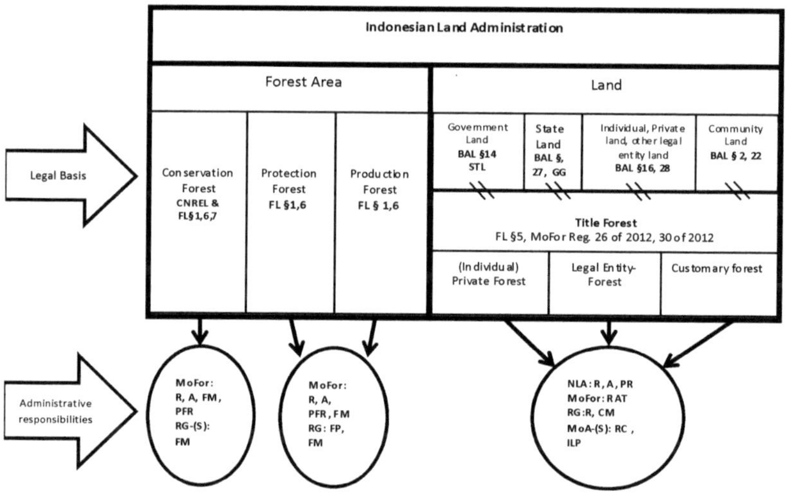

Remark: Secondary (S), Regulator (R), Administrator (A), Provide forest management rights (PFR), Provide land rights (PR), Forest Management (FM), Forest protection (FP), Regulating and administering timber product (RAT), Issue licence for undertaking plantation business (ILP), Regulating agro-commodity (RC), Ministry of Forestry (MoFor), (MoHA), National Land Agency (NLA), Regional Government (RG), Ministry of Agroforestry, (MoA), Article of Law (§), Conservation and Natural Resources and Its Ecosystem Law No. 5 of 1990 (CNREL), Forestry Law 41 of 1999 (FL), Basic Agrarian Law 5 of 1960 (BAL), State Treasury Law 1 of 2004 (ST), Government Regulation 8 of 1953 (GG).

Figure 2 Legal land use transformation options in Indonesia (Article 1)

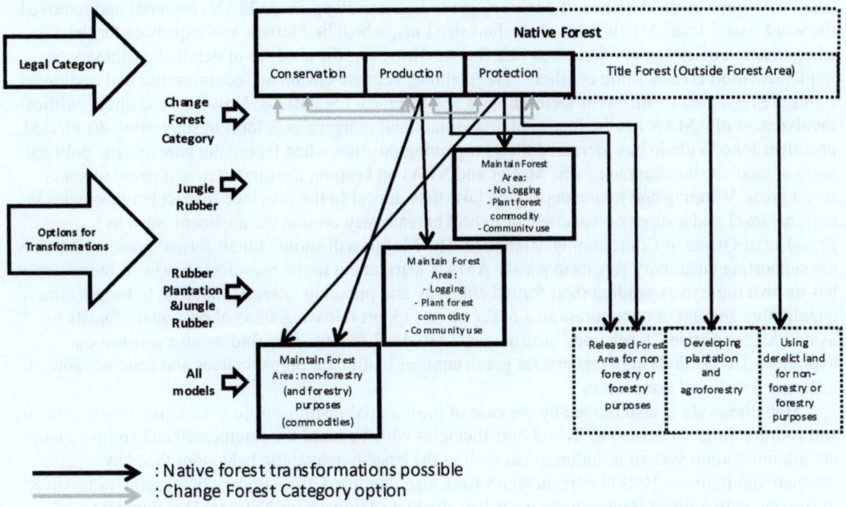

Figure 1 and 2 shows that there are three categories of state forest area: (1) conservation forests, (2) production forests (3) and protection forests. There are also three categories of title forest: (1) private forests (private individual ownership), (2) plantation forests (legal entity ownership) and (3) customary forests. The jurisprudential options and bureaucratic responsibilities are explored in Figure 2 regarding the legal probabilities of the tropical rainforest transformation process.

Figure 2 also confirms hypothesis I, showing that basically each forest category has the possibility of being transformed into any other category in our heuristic cases (agroforestry, conservation, timber plantation, palm oil plantation and rubber plantation) to serve all actual and potential interests of land use change politics and to regulate conflicts between agricultural interests and international and national conservation oriented requirement. To serve the formal requirement of international conservation requirement, land use bureaucracies can allocate state forest for conservation area with high legal basis support. To serve the desire to utilize the palm oil commodity by changing the status of state forests, a legal exception exists if the MoFor, MoA or the regional government follow regulations on utilizing forest areas for non-forestry purposes; so all rainforest models (and commodities) are applicable to forest areas. In spite of this, no forest area has recently been transformed into palm oil plantations. However, practically, it is still illegal to create a palm oil plantation in a state forest. With this, our second hypothesis is confirmed: to serve actual and potential political interests behind the land use transformation system, the Indonesian legal system will provide all jurisprudence possibilities for transforming state forests into non-state forests, particularly providing all legal possibilities to support strong national interests in the palm oil commodity (this national palm oil interest is extensively discussed in Article 3).

Hypothesis I is also proved by our findings on the role of regional bureaucracies on the struggle of indigenous actors fighting for customary forests. The indigenous actors' fight over legal definitions derived by their own interests is also confirmed due to the open legal interpretation reflecting a fragmented and complex authorization system for land use, particularly the definitions of 'forest area' and 'state forest', as a consequence of conflicting interpretations of 'customary forest'. This was also found by segmented socio-jurisprudential research on the Indonesian legal forest administration, as developed in a dissertation by Safitri (2010) from Leiden University in the Netherlands. Our empirical finding showed that Constitutional Court Decision 35 from 2012 later approved *Aliansi Masyarakat Adat Nusantara* or the Alliance of Archipelagic Indigenous People's (AMAN) proposal and removed the word 'state' from Article 1 (6) of the Forestry Law, which had juristic consequences for the categorisation of customary forests as title forests. However, the absence of detailed customary law implementation creates some conflicts in regulations, because customary communities and traditional rights are regulated by different sectorial laws and technical regulations. However, the direct political involvement of AMAN (as the representative of national indigenous actors) in supporting the elected president Joko Widodo has increased their bargaining position when facing the bureaucratic political logic of land use bureaucracies (the MoFor and NLA) on keeping the quantitative amount of state forest areas. Winning and losing depend on how they appeal to the province-district bureaucracies in regional level and a super bureaucracy or hybrid bureaucracy around the president, such as Presidential Office or Coordinating Ministries. The MoFor will show in their formal position that they are supporting customary forests to satisfy AMAN's argument to the president; however, bureaucracy has its own logic in expanding their formal authority and pursuing interests that lead to keeping the quantitative amount of state forest area but seem to support releasing areas of customary forests by using FMU and state CF schemes' instruments (Articles 1 and 6). This finding also confirm our hypothesis IIc that hybrid bureaucracies get in conflict traditional bureaucracies and were not able to influence local land use process.

Hypothesis IIa is also proved by the case of presidential institutions (e.g. task force bureaucracies and coordinating ministries) as hybrid bureaucracies which caused the fragmented and complex land use administration system in Indonesia as well as the basic bureaucratic behaviour faced by international regimes. Hybrid bureaucracies were also identified as secondary bureaucracies involved in steering native forest transformation, such as the State Ministry for National Development Planning/National Development Planning, the Presidential Task Force Unit for Supervision and Management of Development, Geospatial Information Agency, the National Council on Climate Change (NCCC), REDD+ Agency. Hypothesis IIb is confirmed by the result, that REDD+ Agency and NCCC are informally formed to get financial, informational, and regulative support from international regimes.

The president has appointed the Presidential Task Force Unit for Supervision and Management of Development and asked the Corruption Eradication Commission to monitor the memorandum of understanding among the 12 ministries/agencies involved in forest land governance. This indicates that the president is using his coercive power through a super bureaucracy to move beyond the 'traditional agencies' and accelerate the special mission, which is proved our hypothesis IIc. This is in line with Krott (2005), who stated that the main sources of bureaucracies' power of coercion are their legal mandates (including decision rights and sanction mechanisms) and technical resources (enabling bureaucracies to maintain a policy process).

4.2 The Indonesian Sustainable Palm Oil (ISPO): From international governance to the domestic government

One part of hypothesis III is confirmed by our empirical finding that the state imposes obligation to certify. A bundle of regulations from the MoA, the MoFor, the NLA, the Ministry of Environment, and the Ministry of Cooperatives and Small and Medium Enterprises was set up as a legal tool for ISPO standards and principles. The bureaucracy's power (coercive, incentive and informational) has not only led to the successful establishment of ISPO as the new competitor, but also could be used to force voluntary schemes to accept ISPO as complementary to the RSPO standard. This is ambiguous because some ISPO and RSPO standards conflict in the implementation of their principles, especially on legal issues. For example, the NLA is not regarded as being fully supportive of the implementation of High Conservation Value (HCV) from the RSPO standard. Employees of a big palm oil company confirmed that the NLA sent a warning letter to the private company after it applied for HCV in their area. The NLA argued that all land concessions should be managed, otherwise the land might be categorised as wasteland, which is against the law. This not only enforces the reclaiming process but also successfully establishes powerful standards through legal means that affirm the global trade governance mechanism.

In addition, the legality mechanism of regulatory power (Schouten & Glasbergen, 2011) is successful in uniting all power elements from the fragmented Indonesian bureaucracies to support the ISPO (Weber, 1922: 152; Krott et al., 2013). However, this is contrary to Glasbergen's (2011) argument that meta-governance (which is created through private sector and state government interaction) relies in the end on the centrality of states, which comes from the reflection on state roles versus private roles. Our results support his argument that meta-governance is also a highly political process in highly complicated, fragmented networks.

The other part of hypothesis III is also confirmed: global third-party actors were successfully invited by Indonesian bureaucracies to impose cooperation with RSPO-ISPO. The United Nations Development Programme (UNDP) has facilitated the joint collaboration of RSPO-ISPO through joint study audits. However, on the informal position (interest of actors), the joint study audits and even proposals for the formation of single certification schemes are indicative of both schemes assessing the situation and making decisions that benefit both schemes. Indonesian ministries seem to fully support the ISPO while they seem not to oppose the RSPO on a superficial level. RSPO supports ISPO for structural reasons not based on basic interests. These results show that pro-ISPO and pro-RSPO actors formally 'support' each other to shape their informal interests. This phenomenon will lead the RSPO to assert their legitimacy over the ISPO in the future, which is in line with the argument by Overdevest and Zeitlin (2012) that private certification will take back its legitimacy due to the trends of failed public governance and private experimentalism, leading to hypotheses about a strong state-driven race to the bottom benefiting core bureaucracies in Indonesia. In the end, the voluntary and compulsory systems collaborate to attract global initiatives, shaping state and private interests. The outcomes show that there is no 'real' actor opposing either the ISPO or the RSPO from a 'formal position'. This is shown by actors' documents as normative legal arguments. This is in line with Krot's (2005: 8) statement that 'interests are not all openly displayed, but kept secret according to the respective tactics'.

Non-state and international actors seem not to have enough roles in the state mandatory ISPO scheme; however, the RSPO accommodates them (like Sawit Watch and Greenpeace Indonesia, who have good organisational communication with international actors and good access through some farmer groups on the ground). This is in line with Pesqueira and Glasbergen (2013), who note that engagement in non-state governance (RSPO) processes gives NGOs valuable advantages, since they contribute to areas where governments are unwilling to act, and that encompass sector-wide interventions offering a broader potential for change. However, the involvement and backing from NGOs are only supposed to be a 'third pressure' contributed to the market legitimacy to strengthen RSPO's alliance, as explained by Bernstein (2001). However, the ISPO is very confident since all state bureaucracies, as well as some non-state actors like business and farmer associations support ISPO. This 'global-market-legitimisation' will have a new, big challenge from 'domestic-state-legitimation' when the ISPO is introduced to attract powerful 'foreign states' and force them to collaborate. Therefore, international state actors like the EU (and also the United States) seem likely to take a 'wait and see' position toward challenging ISPO management on 'still' providing large numbers and cheap palm oil from Indonesia; this will contribute to the hegemony of neoliberalism (Busch, 2014). Informally, ISPO-RSPO-UNDP cooperation led to many more advantages for RSPO because the challenges of RSPO standard's implementation are now directly mediated with Indonesian core and secondary bureaucracies through UNDP programmes. Hence, this change in governance is opposed by large Western palm oil consuming industries, eNGOs and international actors like the EU. This result supports our third hypothesis that state mandatory schemes supported by state bureaucracies limit the influence of eNGOs and private actors.

Hypothesis IV postulates that the MoA would favour ISPO schemes while the MoT would favour RSPO since the MoA created ISPO and the MoT depends on RSPO's existence for their successful negotiation of international certified palm oil. Surprisingly, our findings that all Indonesian bureaucracies support ISPO rejected this hypothesis. This phenomenon shows that palm oil and ISPO have become a 'national interest' in Indonesian political bureaucracy. 'Reclaiming' is a bureaucratic power process, shaped by actors' interests, where certification authority is reclaimed from private actors. It is inline with bureaucratic interest theory (Niskanen, 1971; Krott, 2005; Peters, 2010) that the interests of bureaucracies as they search for problem-oriented delivery of public services, and their pursuit of the organisational interests of survival and organisational expansion. From the perspective of Indonesian bureaucracies, ISPO appears to be a tool serving the interests of producers, evolved from their bureaucratic achievement based on ISPO expansion and reducing RSPO's power, which only seems to serve the consumers' interests.

4.3 FMU and CF: The utility of national instruments for international and national interests

International utility: Serving international state foreign cooperative actors' missions and sustaining their hegemony

Table 6 shows that there are four main international actors engage in FMU initiatives since 2008 such as GIZ, UK Department for International Development (DFID), World Bank (WB), and Asian Development Bank (ADB). The result shown in Table 6 also proves our hypothesis V that international actors have dual motivations for supporting FMU-CFs: (i) formally they want to find the clearest and efficient way to spend their international cooperation funds in tropical countries and counter global deforestation (ii) informally they want to counter Indonesian palm oil plantations. In addition, this international interest could be contrary to domestic interest in utilizing FMU-CFs. FMU-CF, which is influenced by the international regime, involves both national and international experts.

Political output of FMU which is shown in Table 6 represent high technocratic instrument of forest institutionalizing forest management which can be used to support domestic program and international requirement. Technocratic thinking has produced intricate programs and instruments (such as mechanisms for regulation, budget incentives, information and responsibility), which are very useful for international regimes. These instruments are created using the regulatory powers of the many central bureaucracies involved.

Table 6 Main international actors engaged in FMU initiatives and their projects from 2000-present in Indonesia, and its interests (Article 4)

Actors	Project Title	Duration	Financial Volume	Political output (direct and indirect)	Interests
GTZ/GIZ	Strengthening Management Capacity Programme (SMCP)	2000-2008	No data	- Bringing civil society into FMU policy discussions - Solid FMU policies - Established the National Secretary of FMU development	1. Finding the clearest and efficient way to invest their international cooperation funds in tropical countries 2. Countering global deforestation which means countering Indonesian palm oil expansion
	FORCLIME I	2008-2009	€7.75 million	Experimenting with real FMU models and linking them with global climate change issues	
	FORCLIME II	2013-2016	€14.48 million	Experimenting with real FMU models and linking them with global climate change issues	
DFID	Improving forest management in Indonesia (MFP 3)	2014-2017	£8.8 million	Linking Indonesia's forest certification system with FMU policies and development, especially in terms of tenure and CF issues	
World Bank	FIP	Planning process	World Bank: $17 million	-	1. Finding the clearest and efficient way to invest their international cooperation funds in tropical countries 2. Maintaining the dominance of international fund providers
ADB	FIP	2016-2020	ADB: $17 million Indonesia: $1.5 million	- Assisting a small number of pilot FMUs in becoming operational and engaging with local communities - Stimulating several MoFor agencies related to FMU implementation	

Domestic utility: Multiple utility of FMU-CF instruments

Hypothesis VI is also proved by our finding that domestic bureaucracies used FMU-CF programmes to relocate power back in the central bureaucracies by preparing instruments which formal were in line with international regimes but informal were dominated by domestic bureaucracies. Table 7 shows the interests utilities dominated by domestic bureaucracies such as obtaining financial support, relocating power in the central bureaucracies, countering decentralisation, recentralising information and competencies, and recovering back the state forest area.

Table 7 also provide information on the main and secondary bureaucracies for each interest on underlying FMU-CFs. For example, the MoFor actively promotes sustainable and economic use of forests, as opposed to conservation instruments, which are often seen via their 'technical' characteristics. This means that an instrument solves a political problem at best. BAPPENAS declared that '[without] FMUs, [there would be] no budget' and launched a campaign with the slogan 'No FMUs, no aid' (Setyarso, 2014). BAPPENAS has formally stated that all international donors that invest in Indonesia should directly place their money in FMU development or indirectly situate their programmes at an FMU site or planning area. This is one example of how domestic bureaucracies steer international actors to pursue their interests and prepare welcome direct access for international actors. FMU instruments have shown a strong tendency toward political success. FMUs are not only utilized due to their technical abilities, but also to reflect certain political interests. Many instruments are not introduced due to their 'technical' features with regard to problem solving. Indonesia used a regulatory instrument (which is better in this context than another tool such as a tax or economic instrument); this reflects the government's political interests more clearly. Table 5 summarizes the overall domestic actors' interests, drawn from Articles 4.

Table 7 Summary of domestic actors' interests in underlying FMU-CFs, which is offered to or influenced by international actors and institutions (Article 4)

Utilities	Bureaucracies	
	Main	Secondary
Preparing direct access for international regimes: creating technocratic instruments	MoFor	Ministry of National Planning (BAPPENAS), Ministry of Home Affairs, Ministry of State Apparatus and Bureaucracy Reform
Obtaining international financial support and coalition partners	MoFor	BAPPENAS
Intensify forest utilization	MoFor, Regional Government	Ministry responsible for trade and industries
Countering decentralisation	MoFor	Ministry of Home Affairs,
Recovering state forest areas	MoFor,	Provincial and district governments
Recentralizing competencies	MoFor	-
Recentralizing information and funds	MoFor	-

Experimentation with FMUs is highly successful at the political level in terms of recovering control of state forests and filling the gaps in 'hollow' state forest management. The results showed that there is a strong trend toward recentralisation through the seemingly decentralising act of establishing FMUs (similar case with Mutebi (2004) in Thailand government and Ribot et al. (2006) in forestry sector in several countries). Following the MoFor's efforts to restore its authority and gain complete control over timber licences (Tacconi et al., 2014), a new law from regional governments (Law 23 from 2014) was issued to reduce decentralisation by shifting authority away from district governments to provincial ones. Given this regulatory power, FMUs are now used to enhance the MoFor's efforts to recover state forest areas.

Note on conflicting utility: International versus domestic (Article 4)

Hypothesis V is also proved by three arenas of actual and potential conflicts of interest: 1) among/between international and domestic actors working to counter deforestation at the global level or promote benefits to forests at the domestic level; 2) between various international actors imposing internal agendas; and 3) among domestic bureaucracies in central, provincial and district governments when manifesting FMU as a decentralisation policy or recentralisation strategy.

4.4 The utility of the regional ASEAN land use regime complex for Indonesia

Indonesia is the strongest main actor (along with Malaysia) behind the establishment of ASEAN and has the biggest tropical forests in the region. Indonesia is usually the country that initiates the establishment of regional task forces for specific land use issues under ASEAN, such as forming working groups, expert groups, regional groups and other ad hoc committees. Hypothesis VII was completely proved by our finding that ASEAN has multiple utilities for Indonesian interests as powerful member.

Table 8 shows four utilities of ASEAN land use regimes for Indonesia such as (1) blocking specific rejected international initiatives, (2) attracting funding, (3) imposing standards or rules, and (4) allying against and/or creating a joint selling position against and/or making coalitions with other regimes. On the blocking utilities, Indonesia rejects agreement on *Ramin* timber banned due to its high economic value for Indonesia.

Table 8 The utility of ASEAN forest and land use regimes for Indonesia (Interpreted from Article 2)

Utility	Issues and position	Interest	Instrument used and effort
Blocking	Blocking biodiversity	- Strong interest in palm oil plantation expansion, which is contrary to the biodiversity mission of the secretariat establishment - *Ramin* species is an important and exotic timber in Indonesia	- Not ratifying the biodiversity centre's establishment - Using the ASEAN Expert Groups on CITES and ASEAN Wildlife Enforcement Network to block the Western voice on listing *Ramin* in Appendix I of CITES
	Blocking ASEAN regional timber certification	Indonesia has the strongest national timber certification, which is successfully connected with the EU	- Using Working Group on a Pan ASEAN Timber Certification Initiative to discuss the regional timber certification as the only opinion and initiative at the regional level
Attracting	Attracting funds (not loans)	Linking national initiatives to the regional ASEAN initiatives to attract global funds	Using marketable issues to attract fund, such as: (i) climate change, (ii) regional timber certification, (iii) peatland and mangroves, and (iv) community forestry
Aligning	Common position as ASEAN	Localizing problems and influencing global initiatives before entering the real ASEAN regime and domestic policy	Using regional ASEAN for the common position in the global forum like UNFF and UNFCCC
Imposing	Weakening transboundary haze pollution treaties	Singapore and Malaysia should provide incentives for Indonesian haze pollution reduction	Using ASEAN environmental meeting instruments

In the special case of transboundary haze pollution that affected Malaysia and Singapore, ASEAN successfully developed regional treaties to tackle this issue after 12 years of delays. The agreement is claimed to be the first regional arrangement in the world that binds a group of contiguous states to tackle transboundary haze pollution resulting from land and forest fires. However, Indonesia still imposes minor standards as the coercive binding power. Table 8 also shows that Indonesia utilize ASEAN regimes to attract global fund for specific forest issue finance support or aligning with other member states for the common position in the global forum

4.5 Demonstrating typologies of regime relevance used in domestic politics

One part of hypothesis VIII about a regime-relevant typology is confirmed by quantifying 'problem dimensions' and 'actors' behaviour' using bureaucratic politics. We measured the quality of both dimensions by using quantification and put them into four quadrants of relevance (Figure 4).

As illustrated in Table 9, we used a 0-2-point scale to quantify the indicators of international regimes' scopes and its quality of correspondent with the Indonesia's land use transformation issue and priority for the problem dimension.

Table 9 Indicators for assessing the problem dimension of the relevance of international regimes (Article 2)

Code	Indicator	Scoring Method	Points
P1	Indonesian tropical rainforest transformation issues formally fit in with the regime's scope or aims.	- High match	2
		- Medium match	1
		- No match	0
P2	The problem intensity and or issues quality of Indonesian tropical rainforest transformation system	- High problem	2
		- Medium problem	1
		- Less problem	0
P3	The regime's scope is compatible with Indonesian priority on their forestry transformation's plan	- High priority	2
		- Medium priority	1
		- Little/Not priority	0

Table 10 shows the scoring method on developing actor dimension relevance; by using a scale of 0-2 to rank the indicators of both domestic and international actors on mobilizing resources to intensively utilize international regimes. The actor dimension consist of several elements such as structural in charge of the regimes, budget allocation, legal mandate, and the ability to promote regime and expand coalition.

Table 10 Indicators for assessing the actor dimension of the relevance of international regimes (Article 2)

Code	Indicator	Scoring Method	Points
Domestic actors			
A1	Bureaucratic structures reflecting responsibility for a regime	- There is a specific autonomous bureaucracy dealing with the regime	2
		- There is a specific unit appointed for dealing with the regime, e.g. national focal point, national authority, etc.	1
		- There is no special unit appointed for dealing with the regime	0
A2	Bureaucracies' budget allocations for a regime	- Substantial budget or contribution for additional activities	2
		- Marginal budget allocated according to minimum requirement of a regime	1
		- No budget allocated	0
A3	Bureaucracies' legal mandate to implement a regimes	- High support by laws (*Undang-Undang*)	2
		- Medium support on a legal basis (like government regulation, presidential regulation, presidential instruction) or high support by laws but different interpretation and difficult to implement	1
		- No formal mandate or only tacit agreement within the government	0
A4	Actors promoting regime and broadening coalitions is support for a regime	- Successful coalition building in support of a regime including at least one state bureaucracy	2
		- Successful coalition building in support of a regime among non-sate actor only	1
		- No successful coalition building in support of a regime (minimum one promoter plus one coalition partner)	0
International (and regional) actors			
A5	International bureaucracies' in charge of implementing a regime in Indonesia	- There is a specific unit of an international bureaucracy on regime implementation only in Indonesia	2
		- There is a specific unit of an international bureaucracy on regime implementation in the region, including Indonesia	1
		- There is no specific unit of an international bureaucracy on regime implementation in Indonesia	0
A6	International actors promoting regime and broadening coalitions is support for a regime	- Successful coalition building in support of a regime including at least one state bureaucracy	2
		- Successful coalition building in support of a regime among domestic and international actors	1
		- No successful coalition building in support of a regime (minimum one promoter plus one coalition partner)	0

Hypothesis VIII is fully confirmed as we found four types of regime relevance: 'pure relevance', 'made relevant', 'being made relevant' and 'non-relevant'. Figure 4 shows the relevance typology results from testing our approach on Indonesian land use regime complex cases. Figure 4 shows that the biological diversity, climate change, trade in endangered species, wetland management, international tropical trade, illegal logging, Southeast Asian forest and environmental, the Asia pacific free trade, Southeast Asian rubber trade regime are pure relevant. These regimes mean international regimes address strong domestic problems followed by high support of national bureaucracies.

In contrast, tropical rainforest countries information, Asia and pacific forestry information are concluded as non-relevant (see Figure 4). Hence, although it is part of the international regime complex on rainforest transformation, it currently does not address problems relevant to Indonesia, nor is it useful enough for Indonesian key actors to put their own resources behind it. However, these regimes still exist and provide opportunities for future action if domestic actors deem them useful. For example, shared information regimes play important roles in acquiring or verifying information to maintain Indonesian interests in its tropical rainforest transformation. Information is power, especially verified or pure information, which Indonesian bureaucracies have used in part to enhance their political influence (Simon, 1981; Krott, 1990, 2005; Krott et al., 2013). In this context, this information could be used to oversee the global trends in tropical rainforest transformation. Furthermore, bureaucracies use it to seek advantages in the markets and to balance economic power between tropical forest countries. For example, the MoA tries to collaborate with the voluntary-global palm oil certification to show that their state certification has similar standards and is ready for the global market (Article 1).

Figure 3 Relevance typology of regime complex for tropical rainforest transformation of global and regional regimes in Indonesia (Article 2)

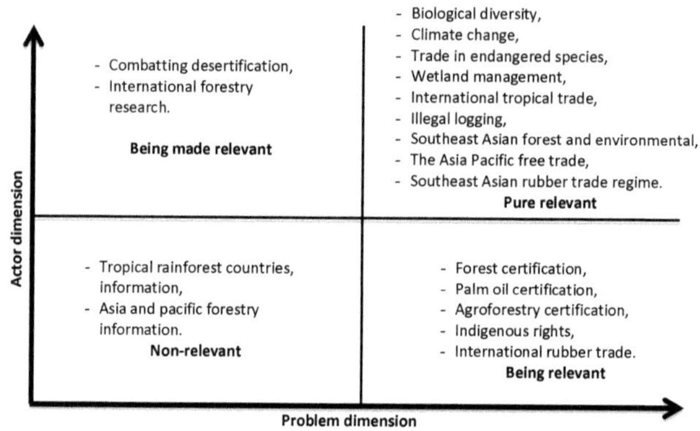

Hypothesis IXa and IXb are also proved as regime types being relevant or being made relevant explains the contrasting interests of Indonesian bureaucracies utilizing global and regional regimes.

Regimes being relevant

The result shows that the forest certification regime, agroforestry certification regime, indigenous rights regime, palm oil certification regime, international rubber trade regime, Southeast Asian forest and environmental regime, and Asia and Pacific forestry information regime were relevant only in the problem dimension. Those regimes with high problem relevance and no actor backing had urgency but were not on the political agenda.

This finding confirms our hypothesis IXa that informally, the bureaucracies oppose the potential solution preferred by the regimes. Special attention is paid to the global rubber trade regime, which is only relevant in the problem dimension. In contrast, the regional Southeast Asian rubber trade regime is relevant in both dimensions. Indonesian actors profit much more from their involvement in the regional rubber trade regime than in the international rubber trade regime due to the fact that in the regional regime Indonesia and Southeast Asian countries can play their roles in a coalition as the largest rubber-producing countries. This suggests that regional regimes have great potential to be of greater use to domestic actors than global regimes, due to a lower number of veto players and shared geographical interests in, for example, tropical forestry (similar to Bhagwati, 2008; Jürging & Giessen, 2013), a phenomenon which might also apply to biodiversity conservation issues in the region.

Regimes being made relevant

Regimes that show high relevance in the actor dimension and low relevance in the problem dimension are not common. However, our result shows that regimes involved in combatting desertification and international forestry research did exist and were relevant only in the actor dimension, not in the problem dimension.

This finding confirms our hypothesis IXb that informally, the national bureaucracies expect to gain advantages from the regimes process. Some forestry policies have been based on the collaboration of policy researchers with these regimes' actors. We found that domestic actors use them to back up their arguments for policy development that fits in with domestic political agendas. For example, the REDD+ Agency or the MoFor may form a partnership with the international research community in order to assess particular programmes and/or policies to back up their forest governance. However, Clark et al. (2002) found that the Consultative Group on International Agricultural Research's research programs that engage different actors are not harming science, but need multiple forms of knowledge to avoid the extreme politicization of formal knowledge.

In addition, domestic action arose during the regimes' involvement due to the specific interests of Indonesian bureaucracies that stood to gain from them; for example, by acquiring more information or verifying information, forming new international coalitions or enhancing existing ones, and even by finding funding sources. The relevance of regimes that have little problem-oriented justification is expected to vanish in the long run. In the short run, however, they have the potential to create substantial political benefits for the actors using them.

5 Strategic options for selected actors

This research contains practical advice for utilizing the ASEAN forest and land use regimes complex. This article arose through a special request from Indonesian official staff at the Centre of Foreign Cooperation of MoFor to provide policy advice for an internal meeting (on 23 March 2015) on preparing for the ASEAN Senior Officials on Forestry (ASOF) institution. Furthermore, the authors have intensive practical experience as co-consultants, working with GIZ to assist ASEAN on providing strategic options for the future timber trade within ASEAN countries. We presented this accomplished consultation work at the 14th Meeting of the Working Group on a Pan ASEAN Timber Certification Initiative, 3-4 June 2015, Manila, the Philippines during our field work.

In addition, the authors, as members of the FKKM, were asked to provide a policy brief for welcoming the new cabinet of the Indonesian Government in the period 2014-2019. The topic was 'challenges and opportunities for CF in Indonesia'. That policy brief on CF including customary forests was published by the FKKM. Our scientific article on CF and FMU (Article 4) is one of the key resources for this practical policy advice.

Therefore, we selected the MoFor delegation in ASOF and indigenous actors as the strategic options that could be considered to deliver their interests. Below, we provide the synopses of those articles and enhance their arguments with our scientific analysis of forestry policy.

5.1 Strategic options for the MoFor delegation on ASEAN forest and land use regimes

5.1.1 Strategy for attracting funding

Indonesia needs to strengthen and expand regional cooperation to attract international funding using ASEAN instruments. Indonesia is advised to maintain its dominant power by hosting the secretariat on specific issues. Competing for host secretariat status in specific institutions among ASEAN countries is driven also by competition for funding.

Germany, the EU, South Korea, Japan and Switzerland continue to be the leading donors to ASEAN regimes, which is relevant for Indonesia when it comes to forest and environmental initiatives and programmes (see Article 1). ASEAN-China, ASEAN Japan and ASEAN+3 constitute strategic co-operation for pursuing these interests. Climate change, regional timber certification, peatland, mangroves and community forestry are currently the most interesting issues for donors. As a newer donor, South Korea is attempting to counter Japan's influence in ASEAN countries and Indonesia, even with regard to Korean and ASEAN collaboration; it aims to build wider cooperation in Asia. South Korean donors direct their funding towards current global issues; typically these donors favour practical issues over policies.

5.1.2 ASEAN regional timber certification: ASEAN Forest Law Enforcement, Governance and Trade Standard

On the supply side, Indonesia, Malaysia and Myanmar are the largest timber producers in Southeast Asia, and they are also important export partners that supply major timber importers such as China, Japan, Korea, Europe, the US and others (Giurca, 2013; Nurrochmat et al., 2014).

Three main factors will influence the effectiveness of the ASEAN Forest Law Enforcement, Governance and Trade (ASEAN-FLEGT): (1) profitability, (2) legality and (3) legitimacy (Nurrochmat et al., 2014). It is very important to make sure that the implementation of ASEAN-FLEGT will increase timber prices for timber exported from ASEAN Member States (AMS), as high costs will be involved with implementing the new legality standard. These costs do not just involve the legal certification following the ASEAN-FLEGT standards, but also the costs of preparation prior to certification, such as administrative procedures and capacity building, as well as costs involved with maintaining certificates in between periodic surveillances (Figure 5).

Hence, the most pragmatic approach would be to adopt a regional mechanism that recognizes the timber legality verification system implemented independently in each AMS. Effort should be made in the report to recommend strategic actions that could assist the AMS in this pursuit. The notion that the ASEAN-FLEGT would result in the dissolution of national systems for the timber trade will inevitably become a stumbling block for the AMS.

Considering the pre-existing condition prior to the ASEAN-FLEGT, then ASEAN through AMAF, ASOF, as well as ARKN (ASEAN regional knowledge network)-FLEGT might do the following:

In the short term:

a. Discuss the need for a legal timber standard as a condition for timber entering into the common ASEAN market, as well as for a non-tariff market, especially for each different AMSs' political and economic interest in an ASEAN timber standard.
b. Indonesia, as it has the most promising national initiatives on sustainable timber legality, needs to continuously facilitate knowledge sharing between member states on initiating domestic legal timber standards to find mutual regional interests on this issue.
c. Assess the principles (criteria and indicators) that can be achieved by member states based on their forestry administration and management contexts and test indicators and criteria for periodic implementation and evaluation.
d. AMAF and/or ASOF through ARKN-FLEGT needs to: (i) describe, justify and assess the risk of compliance and non-compliance with a regional timber legality framework; (ii) identify and develop responsible regional authorities to act as third-party independent auditors to assess the regional timber law framework; (iii) develop a regional mechanism to assess the acceptability of this framework, both acceptable by AMS and international standards; and (iv) develop country-specific control measures for each criterion and indicator in this framework.

In the medium to long term

Three policy options need to be considered if the ASEAN-FLEGT is to be implemented: first, the dissolution of existing AMS national forest governance systems into a single system of ASEAN-FLEGT to avoid inefficient administrative and managerial redundancies; second, whether or not it is better for the ASEAN-FLEGT to serve as an alternative to 'private' forest governance systems and timber legality certifications; and third, whether it is possible to adopt a regional mechanism that recognizes the timber legality verification system implemented independently in each AMS. Effort should be made in the report to recommend strategic actions to be taken to assist the AMS in this pursuit. The notion that the ASEAN-FLEGT would result in the dissolution of national timber trade systems will inevitably become a stumbling block for the AMS.

5.2 Strategic options for indigenous actors

Indigenous actors have split into two contrasting positions: supporting state CF initiatives (such as community forests, people forest plantation, village forests, FMU community forestry and FMU customary forests) or against state CF by asking the central authority to release state forest areas for the customary forest (or non-state CF) or acknowledge customary forest that exist on title land (outside state forest areas). Figure 4 provides strategic options for indigenous actors, whether they are fighting strictly for CF as in the first option or reconsidering backing state CF under specific conditions.

Figure 4 Strategic options for indigenous actors considering state CFs and customary forests

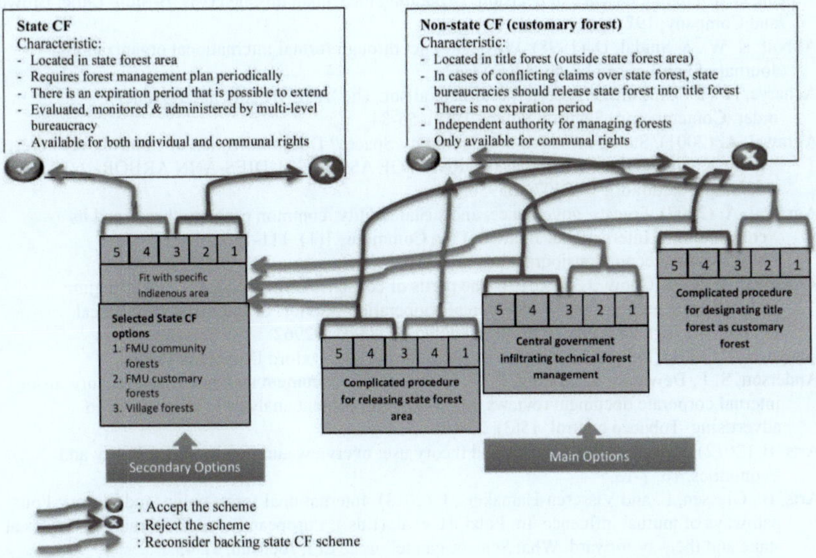

Main option: Fighting for customary forests

In terms of non-state CF (customary forest), indigenous actors (particularly AMAN) have adopted the slogans 'No right, no FMU' and 'No right, no REDD' to fight for legal recognition of their authority and jurisdiction, which now has the strong legal backing from the Constitutional Court. In order to implement the decision, some international donors and global forestry research regimes have been encouraging district governments to issue regulations on customary forests. This is in line with the findings from Article 1, which asserted that it is possible for international regimes to influence domestic actions that help local governments and stakeholders improve forest management and strengthen decentralisation.

Indigenous communities can work with district governments to successfully reclaim customary forests and reject state initiatives, including FMU schemes. An example of this is the customary forest within the conservation FMU. It is noted in Articles 1 and 6 that customary forests have several challenges and conditions, such as: (i) complicated procedures for releasing state forest areas; (ii) central government infiltrating technical forest management; and (iii) complicated procedures for designating title forests as customary forests.

We used a scale from 1 to 5, where 5 showed support for the state CF scheme with specific conditions and 1 showed an inclination to continue fighting for recognition of customary forests.

Secondary option: Reconsidering backing state CF schemes

Based on Articles 1 and 4, state CF schemes such as village forests, FMU community forests and FMU customary forests are the most likely to be considered by indigenous actors if customary forests face difficulty being established. Here we also used a scale from 1 to 5, with 1 suggesting a tendency to fight for customary forests' recognition and 5 suggesting an inclination to go back to fighting for customary forests' recognition.

Reference

Allison, G. T. (1971). Essence of decision: Explaining the Cuban missile crisis. Boston: Little, Brown and Company, 1971

Abbott, K. W., & Snidal, D. (1998). Why states act through formal international organizations. Journal of conflict resolution, 42(1), 3-32

Acharya, A. (1998). Culture, security, multilateralism: The 'ASEAN way'and regional order. Contemporary Security Policy, 19(1), 55-84

Agrawal, A. (2001). State Formation in Community Spaces? Decentralization of Control over Forests in the Kuamon Himalaya, India. JOURNAL OF ASIAN STUDIES-ANN ARBOR-, 60(1), 9-40. http://dx.doi.org/10.2307/2659503

Agrawal, A. (2007). Forests, governance, and sustainability: common property theory and its contributions. International Journal of the Commons, 1(1), 111-136. http://www.thecommonsjournal.org

Aggarwal, V. K., & Chow, J. T. (2010). The perils of consensus: How ASEAN's meta-regime undermines economic and environmental cooperation. Review of International Political Economy, 17(2), 262-290. DOI: 10.1080/09692290903192962

Anaya, S. J. (2004). Indigenous peoples in international law. Oxford University Press.

Anderson, S. J., Dewhirst, T., & Ling, P. M. (2006). Every document and picture tells a story: using internal corporate document reviews, semiotics, and content analysis to assess tobacco advertising. Tobacco control, 15(3), 254-261.

Arts, B. (2012). Forests policy analysis and theory use: overview and trends. Forest Policy and Economics, 16, 7-13.

Arts, B., Giessen, L. and Visseren-Hamakers, I. (2013): International forest policy and Europe: Four pathways of mutual influence. In: Pülzl, H. et al. (Eds.): European Forest Governance – Issues at stake and the way forward. What Science can tell us 2. EFI, Joensuu, 37-43.

Aspinall, E., & Van Klinken, G. (2011). The state and illegality in Indonesia (Vol. 269). Brill.136.

Atkinson, P., & Hammersley, M. (1994). Ethnography and participant observation. Handbook of qualitative research, 1(23), 248-261.

Bachrach, P., & Baratz, M. S. (1963). Decisions and nondecisions: An analytical framework. American political science review, 57(03), 632-642.

Barr, C. M. (2006). Decentralization of Forest Administration in Indonesia: Implications for Forest Sustainability, Economic Development, and Community Livelihoods. CIFOR.

Bakker, L., Moniaga, S., 2010. The Space Between: Land Claims and the Law in Indonesia. Asian Journal of Social Science 38 (2), pp. 187–203. DOI: 10.1163/156853110X490890.

Boyne, G. A. (1998). Public services under New Labour: back to bureaucracy?. Public Money and Management, 18(3), 43-50.

Bowen, G. A. (2009). Document analysis as a qualitative research method. Qualitative research journal, 9(2), 27-40.

Bernstein, S., Cashore, B. (2012). Complex global governance and domestic policies: four pathways of influence. International Afforest areairs, 88(3), 585-604.

Beukema, H., van Noordwijk, M. (2004).Terrestrial pteridophytes as indicators of a forest-like environment in rubber production systems in the lowlands of Jambi, Sumatra. In Agriculture, Ecosystems & Environment 104 (1), pp. 63–73. DOI: 10.1016/j.agee.2004.01.007.

Biermann, F., Siebenhüner, B., Bauer, S., Busch, P. O., Campe, S., Dingwerth, K., & Tarradell, M. (2009). Studying the Influence of International Bureaucracies: A Conceptual Framework. In Biermann, F., Siebenhüner, B (Eds). Managers of Global Change: The Influence of International Environmental Bureaucracies, pp 37-76.

Brockhaus, M., Di Gregorio, M., & Mardiah, S. (2014). Governing the design of national REDD+: An analysis of the power of agency. Forest Policy and Economics. Forest policy and economics, 49, 23-33. http://dx.doi.org/10.1016/j.forpol.2013.07.003.

Brinkerhoff, D. W. (Ed.). (2007). Governance in post-conflict societies: Rebuilding fragile states. Routledge.

Burns, S. L., and Giessen, L. (2014). Identifying the main actors and their positions on international forest policy issues in Argentina. BOSQUE 35(2): 163-173, 2014. DOI: 10.4067/S0717-92002014000200004

Burns, S. L., Yapura, P. F., & Giessen, L. (2016). State actors and international forest certification policy: Coalitions behind FSC and PEFC in federal Argentina. Land Use Policy, 52, 23-29.

Böcher, M. (2012). A theoretical framework for explaining the choice of instruments in environmental policy. Forest Policy and Economics, 16, 14-22.

Comte, I., Colin F., Grunberger, O., Follain, S., Whalen, J. K., Caliman J. P. (2013).Landscape-scale assessment of soil response to long-term organic and mineral fertilizer application in an industrial oil palm plantation, Indonesia. Agriculture, Ecosystems & Environment 169, pp. 58–68. DOI: 10.1016/j.agee.2013.02.010.

Cordoba and Jansen. (2014). The return of the state: Neocollectivism, agrarian politics and images of technological progress in the MAS era in Bolivia. Journal of Agrarian Change, 14(4), 480-500. DOI: 10.1111/joac.12036

Cotton, J. (1999). The" haze" over Southeast Asia: Challenging the ASEAN mode of regional engagement. Pacific Affairs, 331-351.

Cashore, B. (2002). Legitimacy and the privatization of environmental governance: How non–state market–driven (NSMD) governance systems gain rule–making authority. Governance, 15(4), 503-529.

Cashore, B., and Stone, M. W. (2012). Can legality verification rescue global forest governance?: Analyzing the potential of public and private policy intersection to ameliorate forest challenges in Southeast Asia. Forest policy and economics, 18, 13-22. doi:10.1016/j.forpol.2011.12.005

Clark, W. C., Tomich, T. P., van Noordwijk, M., Guston, D., Catacutan, D., Dickson, N. M., & McNie, E. (2011). Boundary work for sustainable development: Natural resource management at the Consultative Group on International Agricultural Research (CGIAR). Proceedings of the National Academy of Sciences, 200900231

Cuyvers, L., De Lombaerde, P., & Verherstraeten, S. (2005). From AFTA towards an ASEAN economic community... and beyond. Centre for ASEAN studies (CAS).

Downs, A. (1967). Inside Bureaucracy. Waveland Press; Boston (292 pp.)

Denzin, N. K. (1970). The Research Act in Sociology. Chicago: Aldine.

Derkx, B., and Glasbergen, P. (2014). Elaborating global private meta-governance: An inventory in the realm of voluntary sustainability standards. Global Environmental Change, 27, 41-50. http://dx.doi.org/10.1016/j.gloenvcha.2014.04.016

Dunleavy, P., and Hood, C. (1994). From old public administration to new public management. Public money & management, 14(3), 9-16

de Jong, W., Arts, B., & Krott, M. (2012). Political theory in forest policy science. Forest policy and Economics, 16, 1-6.

Ebbinghaus, B. (2012). Comparing welfare state regimes: Are typologies an ideal or realistic strategy. ESPAN, Edinburg, UK.

Edwards, P., Giessen, L. (2014): Global forest governance — Discussing legal scholarship from political science perspectives. In: Forest Policy and Economics, 38 (2014), 30-31, http://dx.doi.org/10.1016/j.forpol.2013.08.006

Edwards, D. P., and Laurance, S. G. (2012). Green labelling, sustainability and the expansion of tropical agriculture: critical issues for certification schemes. Biological Conservation, 151(1), 60-64. doi:10.1016/j.biocon.2012.01.017

Edwards P, Kleinschmit D. (2013). Towards a European forest policy — Conflicting courses. Forest Policy and Economics, Volume 33, August 2013, Pages 87-93

Elliott, L. (2000). ASEAN's environmental regime: pursuing sustainability in Southeast Asia. Global Environmental Change, 10(3), 237-240.

Elman, C. (2005). Explanatory typologies in qualitative studies of international politics. International organization, 59(02), 293-326.

Espach, R. 2006. When is Sustainable Forestry Sustainable? The Forest Stewardship Council in Argentina and Brazil. Global Environmental Politics 6: 55-84

Farsund, A. A., Daugbjerg, C., & Langhelle, O. (2015). Food security and trade: reconciling discourses in the Food and Agriculture Organization and the World Trade Organization. Food Security, 7(2), 383-391.
Fenno, R. F. (1986). Observation, context, and sequence in the study of politics. American Political Science Review, 80(01), 3-15
Flatley, W. T., Lafon, C. W., Grissino-Mayer, H. D., & LaForest, L. B. (2015). Changing fire regimes and old-growth forest succession along a topographic gradient in the Great Smoky Mountains. Forest Ecology and Management.
Feintrenie, L., & Levang, P. 2009. Sumatra's rubber agroforests: advent, rise and fall of a sustainable cropping system. Small-Scale Forestry, 8(3), 323-335.
Feintrenie, L., Schwarze, S., Levang, P. (2010). Are Local People Conservationists? Analysis of Transition Dynamics from Agroforests to Monoculture Plantations in Indonesia. Ecology and Society 15(4): 37.
Friedrichs, J. (1990). MethodenempirischerSozialforschung. Westdeutscher Verlag, Opladen, pp. 429.
Giessen, L., Krott, M., & Möllmann, T. (2014). Increasing representation of states by utilitarian as compared to environmental bureaucracies in international forest and forest–environmental policy negotiations. Forest Policy and Economics, 38, 97-104.
Giessen, L., Krott, M. (2009). Forestry Joining Integrated Programmes? A question of willingness, ability and opportunities. Allgemeine Forst- und Jagdzeitung 180: 94-100.
Giessen, Bezerra, Sindt and Jetschke (unpublished). The Rational Design of Regional Forest Regimes: Comparing Amazonian, Central African and Pan-European Forest Cooperation.
Giessen L. (2013). Reviewing the Main Characteristics of the International Forest Regime Complex and Partial Explanation for its Fragmentation. International Forestry Review 15(1):60-70.
Giessen, L., Krott, M., & Möllmann, T. (2014). Increasing representation of states by utilitarian as compared to environmental bureaucracies in international forest and forest–environmental policy negotiations. Forest Policy and Economics, 38, 97-104.
Giurca, A., (2013). Market based and regulatory/enforcement mechanisms- assessment of impacts on timber trade between South-east Asia and Europe, Swedish University of Agricultural Sciences, Master Thesis no, 214 Southern Swedish Forest Research Centre Alnarp 2013,
Glasbergen, P. (2011). Mechanisms of private meta-governance: an analysis of global private governance for sustainable development. International Journal of Strategic Business Alliances, 2(3), 189-206.
Glin, L. C., Oosterveer, P., and Mol, A. P. (2015). Governing the Organic Cocoa Network from Ghana: Towards Hybrid Governance Arrangements?. Journal of Agrarian Change, 15(1), 43-64. DOI: 10.1111/joac.12059
Gourevitch, P. 1978. The second image reversed: the international sources of domestic politics. International Organization, 32(04), 881-912.
Goh, G. (2003). The 'ASEAN Way' Non-Intervention and ASEAN's Role in Conflict Management. Volume 3, Number 1, Spring 2003. Stanford Journal of East Asian Affairs. Pp 113-118
Gomm, R., Hammersley, M., & Foster, P. (Eds.). 2000. Case study method: Key issues, key texts. Sage.
Gulbrandsen, L. H. (2010). Transnational Environmental Governance: The Emergence and Effects of the Certification of Forest and Fisheries. Edward Elgar Publishing.
Guliyev, F. (2011). Personal rule, neopatrimonialism, and regime typologies: integrating Dahlian and Weberian approaches to regime studies. Democratization, 18(3), 575-601.
Halperin, M. H., & Clapp, P. (2006). Bureaucratic politics and foreign policy. Brookings Institution Press.
Hamilton-Hart, N. (2006). Consultants in the Indonesian state: modes of influence and institutional implications. New Political Economy, 11(2), 251-270.
Hatanaka, M., Bain, C., & Busch, L. (2005). Third-party certification in the global agrifood system. Food policy, 30(3), 354-369.
Hellström, E. (2001). Conflict cultures: qualitative comparative analysis of environmental conflicts in forestry (Vol. 2). Finnish Society of Forest Science [and] Finnish Forest Research Institute.
Hemmati, M. (2002). Multi-stakeholder processes for governance and sustainability: beyond deadlock and conflict. Routledge.

Hirsch, P., Warren, C. (1998). Introduction. Through the environmental looking glass. The politics of resources and resistance in Southeast Asia. The politics of the environment in Southeast Asia. Resources and resistance. Routledge, Houndmills, 1-28.

Hovi, J. Sprinz, D., and Underdal, A. (2003). Regime Effectiveness and the Oslo- Potsdam Solution: A Rejoinder to Oran Young. Global Environmental Politics 3:3, August 2003 © 2003 by the Massachusetts Institute of Technology.

Hsieh, H. F., & Shannon, S. E. (2005). Three approaches to qualitative content analysis. Qualitative health research, 15(9), 1277-1288.

Huggins, A. (2015). The Desirability of Depoliticization: Compliance in the International Climate Regime. Transnational Environmental Law, 4(01), 101-124.

Huntington, S. P. (1965). Political development and political decay. World Politics, 17(03), 386-430.

Jones, D. M., & Smith, M. L. (2007). Making process, not progress: ASEAN and the evolving East Asian regional order. International Security, 32(1), 148-184.

Jürging, D.J.J., Giessen, L. (2013). Ein Rechtsverbindliches Abkommen über die Wälder in Europa: stand und Perspektiven aus rechts-und umweltpolitikwissenschaftlicher Sicht. Natur und Recht 35 (5), 317–323.

Kaboolian, L. (1998). The new public management: Challenging the boundaries of the management vs. administration debate. Public Administration Review, 189-193.

Kartodihardjo, H., & Supriono, A. (2000). The impact of sectoral development on natural forest conversion and degradation: The case of timber and tree crop plantations in Indonesia (No. CIFOR Occasional Paper no. 26 (E), p. 14p). CIFOR, Bogor, Indonesia.

Keohane, R. O., & Victor, D. G. (2011). The regime complex for climate change. Perspectives on politics, 9(01), 7-23).

Krasner, S.D. (1982). Structural Causes and Regime Consequences: Regimes as Intervening Variables. International Organization 36(2):185–205.

Krasner, S.D. (1983). International Regimes. Cornell University Press, Ithaca, 372pp.

Krott, M. (1990). Öffentliche Verwaltung im Umweltschutz. Ergebnisse einer behördenorientierten Policy-Analyse am Beispiel Waldschutz, W. Braumüller Verlag, Wien

Krott, M. (2005). Forest policy analysis.Springer, Dordrecht. 323pp.

Krott, M., Bader, A., Schusser, C., Devkota, R., Maryudi, A., Giessen, L., & Aurenhammer, H. (2013). Actor-centred power: The driving force in decentralised community based forest governance. Forest Policy and Economics, 49, 34-42

Lazarus-Black, M. (1994). Contested states: law, hegemony, and resistance. Psychology Press.

Lian, K. K., & Robinson, N. A. (2002). Regional Environmental Governance: Examining the Association of Southeast Asian Nations (ASEAN) Model. Global Environmental Governance: Options and Opportunities.

Lin, L. F., & Zhang, Z. M. (2003). RevisedRank of Rights in Conflicts: An Analysis in the Perspective of Normative Jurisprudence [J]. Journal of Zhejiang University (Humanities and Social Sciences), 6, 000.

Lindstad, B.H., Solberg, B. (2010). Challenges in Determining Effects of international Policy Processes: Forest Protection in Norway as a Case. Forest Policy and Economics 12 (2010) 489-496

Lindstad, B.H. (2014). Changes in Norwegian forest policies: What are the effects of the international forest regime complex?, Scandinavian Journal of Forest Research, DOI: 10.1080/02827581.2014.1000478

Manning, C. (1971). The timber boom with special reference to East Kalimantan. Bulletin of Indonesian economic studies, 7(3), 30-60.

Marin-Burgos, V., Clancy, J. S., and Lovett, J. C., (2014). Contesting legitimacy of voluntary sustainability certification schemes: Valuation languages and power asymmetries in the Roundtable on Sustainable Palm Oil in Colombia.Ecological Economics. http://dx.doi.org/10.1016/j.ecolecon.2014.04.011

Maryudi, A., Nawir, A. A., Permadi, D. B., Purwanto, R. H., Pratiwi, D., Syofi'i, A., & Sumardamto, P. (2015). Complex regulatory frameworks governing private smallholder tree plantations in Gunungkidul District, Indonesia. Forest Policy and Economics.

McCarthy, J., & Zen, Z., 2010. Regulating the Oil Palm Boom: Assessing the Effectiveness of Environmental Governance Approaches to Agro□industrial Pollution in Indonesia. Law & Policy, 32(1), 153-179.
Michon, G., de Foresta, H. (1995). The Indonesian Agroforest Model Forest Resource Management and Biodiversity Conservation. In:Halladay, P., Gilmour, D. A. (Eds), Conserving Biodiversity outside Protected Areas: The Role of Traditional Agro-Ecosystems. IUFRO, pp. 90–106.
McDermott, C.L. 2013. Certification and Equity: Applying an "equity framework" to compare certification schemes across product sectors and scales. Environmental Science and Policy 33: 428-437.
McDermott, C.L., Cashore, B., Kanowski, P., 2010. Global environmental forest policies: An international comparison. Earthscan, London, Washington, DC.
Mosley, L. (2013). Introduction. "Just Talk to People"? Interviews in Contemporary Political Science. In Mosley, L. (Ed.). (2013). Interview Research in Political Science. Cornell University Press.
Müller, H. (2004). Arguing, bargaining and all that: communicative action, rationalist theory and the logic of appropriateness in international relations. European journal of international relations, 10(3), 395-435.
Mutebi, A. M. (2004). Recentralising while decentralising: Centre-local relations and "CEO" governors in Thailand. Asia Pacific Journal of Public Administration,26(1), 33-53. DOI: 10.1080/23276665.2004.10779284
Niskanen, W. (1971). Bureaucracy and Representative Government. Aldine- Atherton, Chicago
Nurrochmat D. R, Dharmawan A. H, Obidzinski K, Dermawan A, Erbaugh J. T. (2014). Contesting national and international forest regimes: Case of timber legality certification for community forests in Central Java, Indonesia. Forest Policy and Economics, In Press, Corrected Proof, Available online 30 September 2014. http://dx.doi.org/10.1016/j.forpol.2014.09.008.
Orsini, A., Morin, J. F., & Young, O. (2013). Regime Complexes: A Buzz, a Boom, or a Boost for Global Governance?. Global Governance: A Review of Multilateralism and International Organizations, 19(1), 27-39.
Osborne, D. (1993). Reinventing government. Public Productivity & Management Review, 349-356.
Ostrom, E. 1999. Self Governance and Forest Resources. Occasional Paper No. 20, Center for International Forestry Research, Bogor, Indonesia. http://www. cgiar.org/cifor.
Overdevest C, Zeitlin J. (2014). Constructing a transnational timber legality assurance regime: Architecture, accomplishments, challenges Forest Policy and Economics, Volume 48, November 2014, Pages 6-15
Pedersen, K. R. (2010). Hybrid bureaucracies? Control and autonomy in small consultancy firms.
Peluso, N. L., & Vandergeest, P. (2001). Genealogies of the political forest and customary rights in Indonesia, Malaysia, and Thailand. The Journal of Asian Studies, 60(03), 761-812.
Peters, B.G. (2001). Politics of bureaucracy. Routledge, London
Peters, B.G. (2010).The Politics of Bureaucracy- An Introduction to Comparative Public Administration.Sixth edition.Routledge. Oxon
Rametsteiner, E., and Simula, M. (2003). Forest certification—an instrument to promote sustainable forest management?. Journal of environmental management, 67(1), 87-98. doi:10.1016/S0301-4797(02)00191-3.
Rahman, Md. S., and Giessen, L. (2014). Mapping international forest-related issues and main actors' positions in Bangladesh. International Forestry Review Vol. 16(6), 2014. 1-16
Raustiala, K., Victor, D.G. (2004). "The Regime Complex for Plant Genetic Resources," International Organization 55: 277–309.
Rayner, J., Buck, A., Katila, P. Eds. 2011. Embracing Complexity: Meeting the Challenges of International Forest Governance. International Union of Forest Research Organizations, Vienna.
Ribot, J. C., Agrawal, A., & Larson, A. M. (2006). Recentralizing while decentralizing: how national governments reappropriate forest resources. World development, 34(11), 1864-1886.
Safitri, M. A., 2010. Forest tenure in Indonesia, the Socio□legal challenges of securing communities' right. Disertasi Doktor. Universitas Leiden, Belanda
Sahide, M. A. K., Burns, S., Wibowo, A., Nurrochmat, D. R., & Giessen, L. (2015a). Towards state hegemony over agricultural certification: from voluntary private to mandatory state regimes on

palm oil in Indonesia. *Jurnal Manajemen Hutan Tropika* [Journal of Tropical Forest Management], Vol. 21, (3): 162-171, December 2015. http://jesl.journal.ipb.ac.id/index.php/jmht/article/view/10798/8499
Sahide, M. A. K., & Giessen, L. (2015). The fragmented land use administration in Indonesia–Analysing bureaucratic responsibilities influencing tropical rainforest transformation systems. *Land Use Policy*, *43*, 96-110. doi:10.1016/j.landusepol.2015.04.030
Sahide, M. A. K., Nurrochmat, D. R., & Giessen, L. (2015b). The regime complex for tropical rainforest transformation: Analysing the relevance of multiple global and regional land use regimes in Indonesia. *Land Use Policy*, *47*, 408-425. doi:10.1016/j.landusepol.2015.04.030
Sahide, M. A. K., Supratman, S., Maryudi, A., & Giessen, L. (2015). Is Indonesia utilising its international partners? The driving forces behind Forest Management Units. Submitted Forest Policy and Economics. (Unpublished)
Schouten, G., Glasbergen, P. 2011. Creating legitimacy in global private governance: The case of the Roundtable on Sustainable Palm Oil. Ecological economics 70(11): 1891-1899.
Setianto, Y. (2010). BIROKRASI TRADISIONAL DI JAWA DALAM PERSPEKTIF SEJARAH. Paramita: Historical Studies Journal, 20(2).
Setyarso, A. 2014. Konsultasi Alignment KPH pada kebijakan dan program Litbanghut. Sekretaris Eksekutif Seknas Pembangunan KPH
Schusser, C., Krott, M., Movuh, M. C. Y., Logmani, J., Devkota, R. R., Maryudi, A., Salla, M., & Bach, N. D. (2015). Powerful stakeholders as drivers of community forestry—Results of an international study. Forest Policy and Economics. doi:10.1016/j.forpol.2015.05.011
Sieniawski, W. (2011). Forest utilization in Indonesia. Sylwan, (4), 279-288.
Simon, H.A. (1981). Entscheidungsverhalten in Organisation. Deutsche Übersetzung, 3. Auflage, Landsberg.
Smouts, M-C. (2008). The issue of an International Forest Regime. International Forestry Review, 10(3):429-432. http://www.bioone.org/doi/full/10.1505/ifor.10.3.429
Sprinz, D. (2005). Regime Effectiveness: The Next Wave of Research. Potsdam Institute for Climate Impact Research & The University of Potsdam.
Stemler, S. (2001). An overview of content analysis. Practical Assessment, Research & Evaluation, 7(17).
Tacconi, L., Obidzinski, K., and Agung, F. (2004). Learning lessons to promote forest certification and control illegal logging in Indonesia. CIFOR.
Terry, L. D. (2005). The thinning of administrative institutions in the hollow state. Administration & Society, 37(4), 426-444. doi: 10.1177/0095399705277136
Thomas, C. W. (1997). Bureaucratic landscapes: Interagency cooperation and the preservation of biodiversity (Vol. 1). Mit Press.
Uszok, A., Bradshaw, J., Jeffers, R., Suri, N., Hayes, P., Breedy, M., ... & Lott, J. (2003, June). KAoS policy and domain services: Toward a description-logic approach to policy representation, deconfliction, and enforcement. In Policies for Distributed Systems and Networks, 2003. Proceedings. POLICY 2003. IEEE 4th International Workshop on (pp. 93-96). IEEE.
Varkkey, H. (2013). Malaysian investors in the Indonesian oil palm plantation sector: home state facilitation and transboundary haze. Asia Pacific Business Review, 19(3), 381-401.
Van Evera, S. (1997). Guide to methods for students of political science. Cornell University Press.
van Noordwijk, M., Agus, F., Dewi, S., Purnomo, H. (2013). Reducing emissions from land use in Indonesia: motivation, policy instruments and expected funding streams. Mitigation and Adaptation Strategies for Global Change, 1-16.
Verd, J. M. (2004). Qualitative research methods.
Wallace, M., & Singer, J. D. (1970). Intergovernmental organization in the global system, 1815–1964: a quantitative description. International Organization, 24(02), 239-287.
Weber, M. (1922). Wirtschaft und Gesellschaft. Tübingen, Germany: Mohr.
Weber, R. P. (1990). Basic Content Analysis, 2nd ed. Newbury Park, CA.
Walter, J. M. N., Torquebiau, E. F. (1997). The geometry of the canopy of a dipterocarp rain forest in Sumatra.Agricultural and forest meteorology, 85(1), 99-115.
Wigell, M. (2008). Mapping 'hybrid regimes': Regime types and concepts in comparative politics. Democratisation, 15(2), 230-250.

Wibowo, A., Sahide, M.A.K, Giessen, L., (2015). Reclaiming legitimacy from voluntary private to mandatory state forest certification: Case of Indonesia. Unpublished

Wibowo, A., Giessen, L. (2012). Identifying international forest-related issues in Indonesia based on the statements in public and expert deliberations. International Journal of Social Forestry. In press

Wibowo, A., & Giessen, L. (2015). Absolute and relative power gains among state agencies in forest-related land use politics: The Ministry of Forestry and its competitors in the REDD+ Programme and the One Map Policy in Indonesia. Land Use Policy, 49, 131-141.

Wicke, B., Sikkema, R., Dornburg, V., Faaij, A. (2011). Exploring land use changes and the role of palm oil production in Indonesia and Malaysia. Land Use Policy 28 (1), pp. 193–206. DOI: 10.1016/j.landusepol.2010.06.001.

Wilson, V. (2014). Research methods: triangulation. Evidence Based Library and Information Practice, 9(1), 74-75.

Wilson, J. Q. (1975). The rise of the bureaucratic state. The Public Interest, 41(3), 77-103.

Varkkey, H. (2012). Patronage politics and natural resources: a historical case study of Southeast Asia and Indonesia. Asian Profile, 40(5), 438-448.

Yong, D. L., & Peh, K. S. H. (2014). South-east Asia's forest fires: blazing the policy trail. Oryx, 1-6. DOI:10.1017/S003060531400088X

Young O.R. (2001). Inferences and Indices: Evaluating the Effectiveness of International Environmental Regimes. Global Environmental Politic tics 1:1, February 201, by the Massachusetts Institute of Technology.

Annex: Constitutive and Additional Publications

Land Use Policy 43 (2015) 96–110

Contents lists available at ScienceDirect

Land Use Policy

ELSEVIER

journal homepage: www.elsevier.com/locate/landusepol

The fragmented land use administration in Indonesia – Analysing bureaucratic responsibilities influencing tropical rainforest transformation systems

Muhammad Alif K. Sahide [a,b,*], Lukas Giessen [a,*]

[a] *Chair of Forest and Nature Conservation Policy, University of Göttingen, Germany*
[b] *Laboratory on Forest Policy and Entrepreneurship, University of Hasanuddin, Indonesia*

ARTICLE INFO

Article history:
Received 23 January 2014
Received in revised form 29 October 2014
Accepted 10 November 2014

Keywords:
Complex national administration
Tropical rainforest transformation systems
Palm oil
Rubber
Formal responsibilities
Informal interests
Bureaucratic politics
Dual system

ABSTRACT

Tropical forests in Indonesia are subject to major transformation processes from native forests to other land uses, including rubber agroforestry as well as rubber and oil palm plantation systems. Using content analysis of policy documents, this paper aims at (i) analysing the formal administrative responsibilities related to the four rainforest transformation systems and (ii) based on the informal motives of the competing bureaucracies involved generating hypotheses on their future course of action and related research. We find that based on the legal and political land use application, Indonesian tropical rainforests may fall into six categories of land use. They may be situated in both, within the forest area and land outside of the forest area in so-called title forests, even though there is a though political debate about forest area category's jurisdiction confirmation. The Ministry of Forestry, the National Land Agency, and regional governments are identified as the core bureaucracies responsible in both forest area and title forest. The Ministry of Agriculture only has responsibilities in title forests. A number of secondary bureaucracies also steering forest transformation are identified. Formal responsibilities of these bureaucracies are highly complex and fragmented regarding the tasks of forest regulation, forest administration, forest management, forest protection, issuing forest management rights, issuing land rights, regulating and administering timber product, issuing licence, and regulating the commodity. Indonesian tropical rainforest is found to potentially transform into other land uses through seven ways: (i) releasing certain area from forest area, (ii) using forest area for non-forestry purposes (e.g. palm oil plantation), (iii) maximising production forests for logging, (iv) developing community forest schemes in forest area, (v) developing plantation and agroforestry in title forest (vi) taking advantage from waste land, and (vii) steering back to native forest. The intersection of responsibilities in steering rainforest transformation has created contestation between the bureaucracies involved. We conclude that the main conflict of interests runs between the core bureaucracies in this transformation, i.e. the Ministry of Forestry, the National Land Agency, and regional governments. The authors conclude with hypotheses on future actions of such bureaucracies in light of national and international influences on rainforest transformation systems. The central hypothesis contends that both core and secondary bureaucracies have conflicting interests over all four types of tropical rainforest transformation due to the areas of overlapping responsibility.

Introduction

Tropical rainforest systems in Indonesia (Steffan-Dewenter et al., 2007), also called native forest are being transformed to other land uses at a high rate (Partohardjono et al., 2005). According to current forest ecology and land use research, these transformations typically follow four recursive patterns, including (i) native rainforest (Walter & Torquebiau, 1997), (ii) jungle rubber (Beukema and van Noordwijk, 2004; Wibawa et al., 2006; Michon and de Foresta, 1995), (iii) rubber plantation (Wibawa et al., 2006; Feintrenie et al., 2010; Michon and de Foresta, 1995), and (iv) oil palm plantation (Wicke et al., 2011; Comte et al., 2013). Each transformation model has its own legal and administrative implications. The legal and political responsibilities of a number of relevant public bureaucracies administering such land use changes are important political factors that can hinder or facilitate the transformation of

* Corresponding author at: Busgenweg 3, 37077 Goettingen, Germany.
Tel.: +49 5513913391; fax: +49 551 39 33415.
E-mail addresses: msahide@gwdg.de, alif.mksr@gmail.com (M.A.K. Sahide), lgiesse@gwdg.de (L. Giessen).

http://dx.doi.org/10.1016/j.landusepol.2014.11.005

rainforests. Hence, it is important to know which state bureaucracy actually is responsible for each rainforest transformation system, including the prior as well as the subsequent type of land use.

Indonesia divides its land into two categories, Forest Area and land outside Forest Area under the legal jurisdiction of the state, but this land systems has raised diverse interpretations (Nurrochmat et al., 2014a,b; Bakker & Moniaga, 2010; Fay & Sirait, 2005). This multiple interpretation is exaggerated and constructed also by such de facto land access activities beyond the formal regulations (Galudra et al., 2013) as well as the challenging the perception of "natural" forest in Southeast Asia (Hunt and Rabett, 2013). Furthermore, the formal responsibilities of different bureaucracies for administering different types of land use have implications on tropical forest transformations. The Ministry of Forestry (MoFor) and National Land Agency (NLA) play indispensable roles in issuing forest land rights (Fay & Sirait, 2005), and thus enabling the legal transformation of Indonesian rainforest systems.

In addition to these two core bureaucracies (MoFor and NLA), it is important to identify other core and secondary administrations along with their formal tasks and responsibilities regarding the four rainforest transformation systems. Hence, this paper aims at (i) analysing the formal administrative responsibilities related to the four rainforest transformation systems and (ii) based on the informal motives of the bureaucracies involved generating hypotheses on their future course of action and related research. Hence, this article addresses the following research questions: which state bureaucracies are legally responsible for legally defined as well as *de facto* forest lands and lands covered by rubber and palm oil plantations.

Methodology

Theoretical underpinnings

Forest tenure and land use administration

Literature on forest and land use administration find many tenure cases on rural and indigenous communities whom have been affected by state policies or the intrusion of outsiders (Larson et al., 2010) as well as state-authorised forest concessions (Anaya & Grossman, 2002). Indonesian state forest categorisation including tenure systems (as well as tree tenure) was originated in the Dutch colonial era and is still strictly enforced (Peluso, 1992; Peluso & Vandergeest, 2001). Furthermore, mapping by government land-use planners focuses on the land itself and efforts of counter-mapping by non-governmental actors to struggle with the nationalistic thrust of Indonesian policy which is emphasised by the homogenising aspect (Peluso, 1995).

On the contemporary forest tenure in developing countries like Indonesia, it involves contestation between the state and civil society (Fitzpatrick, 2006; Wollenberg et al., 2006). The state clearly claims ownership over most forest area (Sunderlin et al., 2009) as the consequence of the post-colonial era which is failed to recover forest dwellers rights (Peluso, 1995; Movuh and Mbolo, 2012). However, Indonesia has successfully developed regional decentralisation including on forestry sector and found that local social forces expanded more influence and performed to limit the local government (Anderson, 2000; Nurrochmat, 2005; Wollenberg et al., 2006).

Bureaucratic politics explaining land use administration

Weber's theory on bureaucracy defines bureaucracy as a system of organising action with six main characteristics: (i) rules are the basis for action (laws, regulations, etc.) in either public or private life, (ii) offices for the management of rules are hierarchically organised and supervised, (iii) knowledge of action is contained in records, not in individuals, (iv) specialised training and expert knowledge is required to perform offices, (v) official business is the main activity of individuals holding office, and (vi) management of the offices is governed by learnable rules.

'Bureaucracy steering' is the actor-centred analysis of identifying who is charged with both formal responsibilities (such as legal procedures and specific tasks) and 'informal mission', or interest. In this paper the argumentation is founded on bureaucratic political theory, which states that bureaucracies have distinct formal tasks for delivering public services (in this case concerning the forest land use administration and management), which are their outputs (Niskanen, 1971); as well as informal tasks, like competition for jurisdiction over land with other bureaucracies, for resources, political domains (e.g. REDD+, forest moratorium, illegal logging), and influence (Allison, 1971; Niskanen, 1971; Stern, 1998; Krott, 1990; Hubo & Krott, 2010; Peters, 2010). Consequently, bureaucracies have two main goals; they strive for problem-oriented delivery of public service as publicly stated in their mandates, and they pursue the organisational interests of survival and expansion (similar Giessen et al., 2014; Giessen, 2011). On the simple organisational interest category, bureaucracies can be divided into two orientation based on their basic interest. First, production oriented bureaucracies and the second is conservation oriented bureaucracies (Hirsch & Warren, 1998). This is in line with the argument that in cases where both interests cannot be pursued simultaneously, organisational interests are given higher priority (Niskanen, 1971; Krott, 1990; Peters, 2010).

Political influence manifests in the ability to formulate or implement policies adherent to one's own interests. Influence is exerted by the use of information and power (Simon, 1981; Krott, 1990, 2005). *Power* is the ability to shape a programme according to one's owns interests even against resistance from other actors. At the core of this concept lies Weber's definition of power as the "probability that one actor within a social relationship will be in a position to carry out one's own will despite resistance" (1922, 152). In this paper, we will focus on the coercive element of power only, by looking at the formal responsibilities, based on which a specific bureaucracy can grant permissions (e.g. release land from forest area) or prohibit things.

Methods

Friedrichs's content analysis of formal policy documents (1990) is applied. This content analysis is based on our expert experiences through two steps. The first, we select the most relevant and the most current policy document including national constitution, laws, constitutional court decisions, government regulations, presidential regulation, presidential decree, ministerial regulation, ministerial decree, and Memoranda of Understanding. The second, we analyse the documents by filtering out all statement relevant for our research question that explain clear formal tasks and responsibilities. This analysis is applied to both Basic Agrarian Law (BAL) and Forestry Law (FL), and additionally to all laws and regulatory documents related to the dual system of forest land administration. These legal documents are taken from official government websites or other trusted sites. In addition, founding documents of institutions, programmatic documents, and other archival sources that articulate the formal responsibilities of the core and secondary bureaucracies are cited. In Part 3.4, jurisprudential options and bureaucratic responsibilities are explored on the legal probabilities of the tropical rainforest land use transformation process. In Part 4, hypotheses are presented on the informal goals of bureaucracies administering land uses and transformation by using bureaucratic politics theory.

Results and discussion

Forest-related land use administration in Indonesia within a dual system

Land allocation processes in Indonesia are a political process (Nurrochmat et al., 2012) and have been ruled by conflicting policies (Brockhaus et al., 2012). Mainstream policy divides Indonesian land into two categories: *Kawasan hutan* which we define as forest area, and all other land outside forest area. Both land regimes are under state control, regulated mainly by the *Basic Agrarian Law* (BAL), which is also referred to as *UUPA* (*undang-undang pokok agraria*) 60 of 1960, and the Forestry Law (FL), which is also referred to as *Undang-Undang Kehutanan* 41 of 1999, and both of which govern land use (Fay & Sirait, 2005; Bakker & Moniaga, 2010). Forest area is categorised based on its fulfilment of one of three functions, which includes (i) production forests, (ii) protection forests, and (iii) conservation forests. In the Indonesian case, forest area is state granted or designated by law. Forestry Law 41 of 1999 Article 1 paragraph (3) states that forest area is "any particular area determined or designated by the government to be permanent forest". Andiko (2010) argues that these imprecise definitions can lead to land without standing trees being gazetted as forest area. Furthermore, there is no final jurisdiction status confirmation of *kawasan hutan* definition especially on the state ownership interpretation, but the term of 'designated by government' and 'permanent forest' has been interpreted by rules on the ground (Government Regulation, Ministry Regulations, and even other sectoral Laws like Spatial Planning Law) that *kawasan hutan* or forest area is managed by state under MoFor authority and treat it as 'state owned'. However, some policy experts and practitioners stick to *kawasan hutan* definition that all forest land use (including non-state forest) should be categorised under this term or forest area. In our analytic point of view, we categorised this forest land use based on the legal and also political application in the field. Therefore, we exclude non state forest from forest area.

This paper differentiates non-forest area in so-called 'title forests' into at least three main forest land use categories such as private forests (represent individual ownership), customary forests (represent indigenous community ownership), and legal entity forest (represent company or legal institute ownership). These categories are the recent policy development after Constitutional Court Decision 35 of 2012 which is explained in Section 'Administration of title forests.' Hence, MoFor is formally responsible for approximately 61% of Indonesian territory classified as forest area, while NLA is formally responsible for approximately 39% classified as non-forest area (Fay & Sirait, 2005). Nevertheless, there is a cross-responsibility intersection between both organisations. Title forest is the only legal category of forest outside forest area, and though the term 'title forest' does not appear in BAL, it is contained in Forestry Law 41 of 1999 and regulations under the jurisdiction of Forestry Law. In Article 3 of MoFor Regulation P.30 of 2012, it is clearly stated that title forest is established by (a) certificate of ownership, Letter C or *Girik* (a notification letter from either Dutch colonial era or customary authority), (b) *HGU* certificate or right of use licence, or (c) other NLA documents that confirm land ownership or land occupation outside forest area. Title forest land is inclusive of a bundle of rights under BAL, but which is exclusive of the forest itself. However, all legal land status is not clearly mentioned on BAL, but some rules make an interpretation such as Government Regulation 8 of 1953 on Mastery of State Land for explaining 'state land' and State Treasury Law 1 of 2004 on describing 'government land' (see also Fig. 1).

MoFor has responsibilities to administer and govern the legality of forest products (especially timber) from title forest, which has an additional bundle of rights issued by NLA under BAL. There are effectively two bundles of rights on title forest; the first is granted by NLA, while the second is registered by MoFor for forest resources (timber) grown on that land. MoA as production oriented bureaucracies is responsible for agroforestry products, especially for plantation commodities outside forest area. This bureaucracy has developed a special unit for plantation issues called the 'Directorate General on Estate' under MoA. Cross-responsibility intersects for both commodity management and natural resource management, both of which have entrenched interest-based motivations.

In the scientific literature, the terminology of tropical rainforest is ambiguous due to the lack of a precise definition (Swaine & Whitmore, 1988). In this paper, the 'ecological law' definition from Forestry Law 41 of 1999 applies, which describes forests as "an ecosystem unit in term of a plot of land containing bio-natural resources dominated by vegetation in integrated unity of environment thereof" (Art. 1, paragraph 2). Based on this legal definition, Indonesian forest cover could be interpreted to include all categories of land (both forest area and non-forest area). The idea of state sovereignty over a category of land cover called "Forestry" emerged in South-East Asian after World War II, creating powerful institutions of state forestry (Peluso & Vandergeest, 2001).

During political transitions (into decentralisation), Indonesian country is particularly vulnerable to the fragmented government into provincial and district level (Smith et al., 2003). Regional governments in autonomous regions[1] (district government or provincial government) have been partially mandated with delivering services in land administration under Law 32 of 2004 on Regional Government. NLA is otherwise the main administrator on the land outside forest area. An intersection of responsibilities occurs in rights licensing power and interest-based jurisprudential administrative responsibility. The consequence of this rights overlapping produces contestation between MoFor, NLA, and regional government over the "middle area" which exists both in policy formulation and in service delivery. One example is the (public) need of land for development infrastructure (housing, infrastructure, public offices building), especially for the new districts in Indonesia, then Regional Government can use Spatial Planning Law (Law No. 26 of 2007) to revise their regional allocation plans and transform forest area into non forestry purposes. Non Forested Area or APL can be interpreting as the middle area. APL is not forest area but on the real landscape it is possible non-forest and also forest. APL can be transformed into forest area through the land swap mechanism (this can be seen in Section 'Released forest area and change forest category').

On 21 February 2012, the Constitutional Court granted the plaintiffs' request through Constitutional Court Decision 45 of 2011, declaring the phrase "designated and or" in the Article 1 (3) of Forestry Law to be unconstitutional and unenforceable. Phrase designated and or gazetted is deleted and change with only phrase "gazetted" (Indonesian Constitutional Court, 2011). Article 1 (3) now reads: "Forest Area (*Kawasan hutan*) is a particular area designated (*ditunjuk*) and or gazetted (*ditetapkan*) by the Government to be maintained as permanent forest (*Hutan Tetap*)." Furthermore, MoFor Regulation 44 of 2012 that the gazette of forest area required final forest boundaries which is very difficult to implement in the field due to forest land tenure conflicts among the stakeholders.

[1] Based on the Regional Government Law 32 of 2004, we conceptualise autonomous government as the devolution power to either provincial government or district level in the context of decentralisation. However, autonomous government is mostly relying on the district government (meanwhile, province government is vested with only limited autonomy). But in the forestry case, provincial government take responsible on inter-districts based forest management and administration.

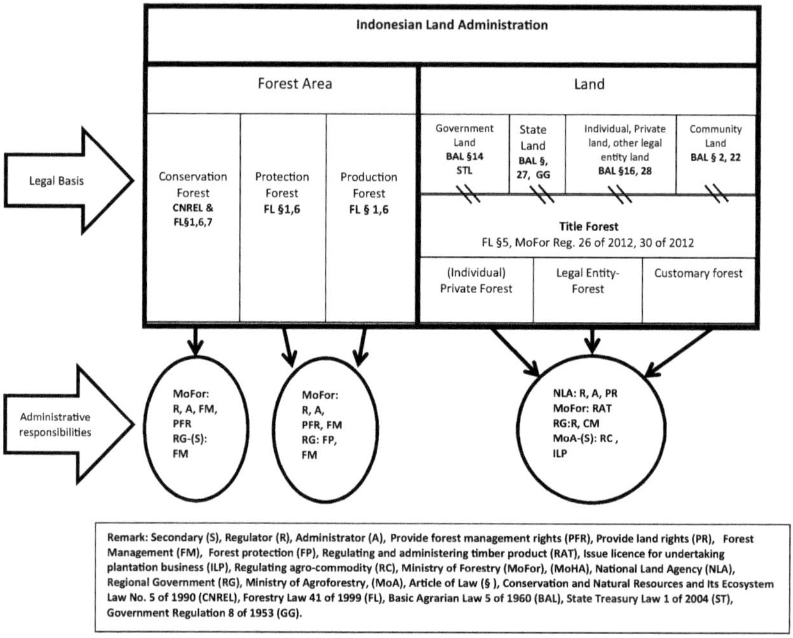

Fig. 1. Dual system of Indonesian land administration.

The implications of this ruling are potentially thoughtful concerning (i) the recent legal status of forest area, (ii) the ability of the MoFor to exert management authority over it, and (iii) changes to the formal and informal balance of power between central and regional government authorities in determining the allocation of land to forestry versus non-forestry purposes within provincial spatial plans (Wells et al., 2012; Arizona et al., 2012; van Noordwijk et al., 2013). Meanwhile, MoFor argued that this Constitutional Court 45 of 2011 is not changing anything, it only has implication on the future of forest land use administration which should be gazetted due to the principle of retrospectivity on Constitutional Court jurisprudence or its decisions cannot affect individual rights or legal relationships established in the past (Arizona, 2013; Wells et al., 2012). This fragmented dual land system can be seen in Fig. 1 and the details explanation will be explored in Section 'Administration of title forests.'

Administration of title forests

Category of title forest

NLA holds official jurisdiction and organisational responsibility for all types of rights registered at the state level. Meanwhile, we differentiate title forest into three forest forms based on the rights in BAL, and also Article 1 of Forestry Law stated 'by Title Forest we mean any forest situated on land charged with land title'. 'Title' on this Forestry Law is abundant of land rights outside forest area (MoFor Regulation P.30 of 2012). These include private forest under right of ownership, forest plantation under right of cultivation (*Hak Guna Usaha/HGU*) and customary forest under indigenous right of tenure. This is also strengthened by Constitutional Court 35 of 2012 that state forest (forest area) as referred to in paragraph (1) point a. does not include customary forest. There are only two categories in Forestry Law (Article 5) paragraph (1) "Forest shall by status consist of: a. state forest, and b. title forest. Hence the category of title forest within should include customary forest (see Section 'Forest-related land use administration in Indonesia within a dual system'). Hence, BAL, Forestry Law and Constitutional Court Decision correspond each other to describe our title forest categories below.

a. **Individual – private forest** is commonly defined as *hutan rakyat*, or people forest, and is attached to the right of ownership or to any rights that authorise forest farmers to manage their forests (like right of use and right of lease). In this term, we classify private forests as the individual (forest farmer). The amount of private forest is about 1.568.415,63 ha (MoFor, 2005).
b. **Legal entity – forest** is categorised as any agricultural activity, including rubber and palm oil plantation under the right of cultivation, or business permit (*HGU*) terms of up to 35 years that can be extended for additional 25 years. Individuals and firms (both state firms and private companies) are eligible to access an HGU certificate as long as they fit requirements including Indonesian citizenship, and for firms, incorporation and domicile in Indonesia based on the Article 30 of BAL and Article 2 of Government Regulation No. 40 of 1996. Actually, forest plantation is not a common term used on forestry especially on the title forest due to the fact that the plantation sector in title forest is controlled and regulated under MoA. MoFor will only regulate the logging activity of plantations which are located out of forest area. Rubber plantations covered about 3,107,544 million ha in 2013 (MoA, 2013).

c. **Customary forest** is forest managed by indigenous people under *hak ulayat* (indigenous) or other similar rights of *adat* (custom law communities) as long as such communities in reality still exist. It is consistent with state interests and shall not contradict or supersede other laws and regulations (Arts.2 and 3, BAL). These rights are strengthened by Indonesian Constitutional Court Decision 35 of 2012. There is no final confirmation formal data about the amount of this type of forest cover.

Recent legal developments on customary forest

According to the National Constitution, 1945 article 18B (2), the state recognises and respects customary communities and their traditional rights based on the law. Sirait et al. (2001) have identified laws and regulations related to indigenous rights in as many as seven Laws (including BAL and Forestry Law), and also in one People's Consultative Assembly's Resolution (TAP MPR), two government regulations, one presidential decision, one presidential instruction, three ministerial regulations, and four ministerial decisions. With these laws and regulations approved the possibilities of customary forest (as part of customary rights) are found both on forest area and outside forest area. Even though there are so many regulations about respecting indigenous land rights as established by the principle of customary forest, there is no detailed regulation about how to implement the establishment of customary forest (especially) categorised as forest area based on FL. Constitutional Court Decision 35 of 2012 later approved *Aliansi Masyarakat Adat Nusantara* or the Alliance of Archipelagic Indigenous People (AMAN) proposal and removed the word "state" from Article 1 (6) of the Forestry Law, which has juristic consequences for the categorisation of customary forest as title forest.

Government Regulation 10 of 2010 on Forest Determination and Function Change Procedure (Art. 2) offers re-leasing of forest area based on "development dynamic and community's aspiration". This language gives bureaucracies a way to implement the Constitutional Court Decision 35 of 2012, while also stipulating that forest area relinquishment should maintain forest area proportions of 30% of province-wide area or watershed area. This proportional requirement may potentially oppose AMAN's claim that there are at least 40 million hectares of customary forest in Indonesia, or almost 50% of existing forest area (AMAN, 2013). Furthermore, NLA's role is conspicuously absent in this regulation. Other potentially puzzling issues arise because BAL (Arts.2 & 3) requires that the 'establishment of customary forest should not be contrary to the national interest'. This allows for political steering of native forest transformation with an open interpretation based on the variety of bureaucratic interests on customary forest issues.

The director of NLA agrees with Constitutional Court Decision No. 35 of 2012 (HuMa, 2013), however, this legal decision is a merely conditional recognition that customary forests are those already legally recognised by Article 67 of FL, which states that claims should be legalised through regional government regulation which attach clear boundary and institutions. Soon after this Constitutional Court decision, AMAN reported that many indigenous groups began ad hoc mapping of their customary forest claims with stick boundary markers. The recent MoFor regulation has been issued (MoFor Regulation P.62 of 2013 on the Amendment of MoFor Regulation P.44 of 2012 on the Inaugural of forest area) which is including Constitutional Court decision No. 35 of 2012. These regulations wait for the regional government response whether they recognise indigenous by issuing regional regulation.

Village Law 6 of 2014 has clearly mentioned the possibility of forming customary institutions in village levels and as part of formal village institutions (Article 95). By this law, Regional Government could possible form Customary Village. But this law is not clearly stated about customary court which is the essence of strengthening customary and traditional rights. The absence of customary law creates some conflicts of regulations because customary communities and traditional rights are regulated by different sectorial laws and technical regulations. AMAN as customary representative takes neutral positions of this Village Law and still proposes Recognizing Indigenous Rights Law in the parliament.

Bureaucracies responsible for title forest

Main bureaucracies. NLA is the main administrator of Indonesian land outside forest area. It has led government in the recognition and granting of seven types of land rights and three types of natural resource rights. *Hak milik* (right of ownership) is considered the most powerful right in Article 16 of BAL. The other rights clearly stated in BAL such as (i) *hak guna-usaha* (right of cultivation), (ii) *hak guna-bangunan* (right of use of structures), (iii) *hak pakai* (right of use), (iv) *hak sewa* (right of lease), (v) *hak membuka tanah* (right to clear land), (vi) *hak memungut-hasil-hutan* (right to collect forest produce), and (vi) rights other than those mentioned above, which shall be stipulated by way of an Act, and rights of provisional nature. Not all of these rights have been parsed into detailed regulation. However, Indonesia's decentralised political system (Regional Government Law 32 of 2004) mandates district and provincial governments to act in a supporting role to NLA in land services. Central government commands the issuance of main land rights, management, policy making, and regulatory aspects. Table 1 illustrates regional government's role in mediation of land conflict, its provision of preliminary licences for location permitting, and permission to open land for the procedural process to obtain Right of Cultivation from NLA.

MoFor issued Regulation P.30 of 2012on the administration of title forest products, which also allocates the responsibilities and roles of regional government (both provincial and district governments). MoFor is the core bureaucracy of Forestry Law, and therefore is responsible for administering and governing the legality of production in title forest and to ensure its ecosystem preservation. MoFor Regulation No P.26 of 2005 (Art. 15, paragraph 2) determines that title forests can produce timber and non-timber products, as well as products of ecosystem preservation. This regulation has also been enhanced by MoFor Regulation P.30 of 2012 on forest products (specifically timber production) from title forest. Regulation P.30 of 2012 regulates only forest timber products, not land rights. NLA, as the core bureaucracy of BAL, registers land rights. Table 1 provides a brief overview of various bureaucratic responsibilities.

Secondary bureaucracies. Secondary bureaucracies can be ministries, task forces, or inter-ministerial bureaucracies dispatched with special tasks that are entangled among related ministries.

Rubber and palm oil plantations are mostly located in title forest. Both commodities are prioritised in the Master Plan for Acceleration and Expansion of Indonesia's Economic Development for 2011–2025 on Sumatra Island (SMNDP, 2011). According to the National Planning System Law 25 of 2004, the State Minister for National Development Planning/National Development Planning (SMNDP) is responsible for providing planning for rubber and palm oil plantations. Under the same law, the regional government is responsible for providing regional development planning (both on provincial and district/municipality level). Law 26 of 2007 on Spatial Planning grants SMNDP and autonomous regional governments shared responsibilities in arranging the spatial planning for rubber and oil plantation in title forests. The Ministry of Trade (MoT) is responsible for increasing trade in palm oil and rubber, most of which is planted on title forest. MoT is also responsible for promoting palm oil commodities as the green product on international trade forums such as APEC. Meanwhile, Ministry of Industry is (MoI)

Table 1
Main bureaucracies' responsible for title forests administration.

	NLA	Regional Government (District Government)	MoFor	MoA
General Responsibilities	Administer and regulate land rights (Based on Regulation of the Minister of Agrarian Affairs/Head of NLA 3 of 1999) Temporary task: Delay the issuance of land rights (for HGU and others) on the moratorium area for the period 2013–2015 (Based on Presidential Instruction 6 of 2013);	Provide services on land administration, such as use of vacant land and settlement issues; determines subject and object-land redistribution; guarantees, replacement of excess soil loss; and planning of land use at district/municipality level (Stated at the attachment of Government Regulation No. 38 of 2007)	Regulate timber product administration and legality (Based on MoFor Regulation P.30 of 2012) Temporary task: Delay the issuance of any management rights; recommendation and location permit for logging on the moratorium area for the period 2013–2015 (Based on Presidential Instruction 6 of 2013.	If land is managed by agroforestry systems that produce agricultural products, MoA regulates the agricultural products
Specific Responsibilities on Private Forests	Grant property rights to agricultural and non-agricultural land, grant the right to use agricultural and non-agricultural land (Based on Regulation of the Minister of Agrarian Affairs/Head of NLA 3 of 1999)	→Administer timber transportation documentation (Based on MoFor Regulation P.30 of2012) →Implementation of land acquisition for development purposes; dispute resolution; completion of restitution and compensation issues for land development (Stated at the attachment of Government Regulation 38 of 2007)	Regulation of forest products from private forest.	Write regulation and incentives for agricultural production (from agroforestry model on private forest).
Specific Responsibilities on *HGU* Plantation Forests	→Grants leasehold or business permits (*HGU*) for plantation (Based on the Government Regulation 40 of 1996)	→Administer timber transportation documentation (Based on MoFor Regulation P.30 of 2012) →Issues location permits, grant permission to open land (Based on Regulation of the Minister of Agrarian Affairs/Head of NLA 3 of 1999), Facilitates Environmental Impact Analyses (together with central government) (Government Regulation 27 of 1999)	→Regulate and issue the relinquishment of forest area (for plantation)	→Issue licence for undertaking plantation business (IUP) (Based on the Ministry of Agriculture Decision No. 786 of 1996). Issues regulation on the guideline of Plantation business (Ministry of Agriculture Regulation 98 of 2013)
Specific Responsibilities on Customary Forests	Register customary forest	→Verifying customary forest land and institution, providing District Regulation (Based on the BAL and partly explained on the attachment of Government Regulation 38 of 2007)	→Administer forest products	

responsible for expanding markets in both domestic and international trade and for spreading industry outside Java Island.

Indonesia's Transmigration Programme began during Dutch colonial era in the early 1900s, and it plays an indispensable role in palm oil and rubber plantations development in Sumatra (Zen et al., 2005; Casson, 1999; Budiman & Penot, 1997). In the beginning, this programme mobilised labour from Java and Bali Island to estate area development in rubber, oil palm and cocoa. It applied Nucleus Estate Scheme (NES) to facilitate partnerships between big companies (both state company and private company), district government, and labour to provide housing, seeds, implements, and two hectares of land for the migrant labour. The Ministry of Manpower and Transmigration (MoMT) is responsible for coordinating a district-to-district basis transmigration programme and for coordinating costs shared between the central office, the districts involved, and the plantation beneficiaries. Regional government is responsible for the management of migration and transmigration settlements, and to provide guidance for field work, facilitate business land acquisition, establish residence with the status of land ownership, and guidance, development, and protection of business partnerships (Based on Law 29 of 2009 on the Amendment of Transmigration Law 15 of 1997). The palm oil product evolution into national biofuel programme has involved the Ministry of Energy and Mineral Resources (MoEMR) as one of the main agency on providing incentives and facilitating regulation (MoEMR, 2013).

Customary forests have been drawn up with assistance from indigenous peoples acros the country and facilitated by the Participative Mapping Working Network (JKPP). JKPP (2013) reports that work on customary forests borders found 2.4–2.6 million hectares of customary forest in Indonesia. Its findings have been sent to the Presidential Task Force Unit for Supervision and Management of Development (UKP-PPP), Ministry of the Environment (MoE), and Geospatial Information Agency (GIA). Several MoA and cooperation charters have been signed between AMAN and the National Commission onHuman Rights (NCoHR), and between AMAN and Minister of Environment (AMAN, 2013), to mediate conflict in customary forests establishment.

UKP-PPP is working closely with presidential institutions to supervise and coordinate ministries and local governments to run national development (based on the Presidential Regulation 54 of 2009 and Presidential Regulation 10 of 2012). UKP-PPP monitors the process of establishing customary forest pursuant to Constitutional Court Decision 35 of 2012 (Satgas REDD+, 2013a). UKP-PPP also monitors and evaluates applicable international and bilateral forest regimes such as REDD+ and forest moratorium (UKP-PPP, 2013). The moratorium is a programme between the governments of Indonesia and Norway that delays issuance of land rights (for HGU and others) on the moratorium area from 2013 to 2015 (as also shown on Table 1). Even though the Ministry of Foreign Affairs (MoFA) holds competence in the international relations dimension, other related bureaucracies are involved in Indonesian lobby representation in international negotiations forums of forestry regimes. Climate change regimes represent a prime example. As a result, the president created the National Council on Climate Change (NCCC), through Indonesian Presidential Regulation Number 46 of 2008. NCCC is responsible for coordinating the implementation of climate change policy and for strengthening Indonesia's position in international forums on climate change control.

Bureaucracies responsible for forest area

Forest area has been classified based on three functions: (i) conservation, (ii) protection, and (iii) production.

a. **Protection forests** are those designated to protect, a buffer that secures the water system, prevents floods, controls erosion, prevents sea water intrusion and maintains soil fertility. The total coverage area is about 32,211,814.72 ha (MoFor, 2012a)
b. **Production forests** are those designated to yield forest products. In Regulation P.50 of 2009, MoFor asserts that production forest falls under three categories, including (i) Convertible Production Forests, (ii) Permanent Production Forest, and (iii) Limited Production Forest. The total coverage area is about 77,835,293.99 ha (MoFor, 2012a)
c. **Conservation forests** are those with specific attributes functioning to preserve flora and fauna diversity and ecosystems. This total cover area is about 8,025,820 which is not included in water conservation areas (Ministry of Forestry, 2012a)

Main bureaucracies

Here, responsibilities of bureaucracies are divided into two categories: (i) forest administration and (ii) forest management. The scope of forest administration are regulation, general planning, and granting rights. Forest management refers to more specific on-site responsibilities, such as forest planning, forest inventory, and forest management for income generation. This broad responsibility classification is illustrated in Table 2.

Table 2 illustrates that the arrangement of formal responsibilities in forest administration and forest management is distributed between both central and regional government due to the forest decentralisation which is under the jurisdiction of Regional Government Law 32 of 2004. A temporary additional task of these bureaucracies is to delay the issuance of land rights (for HGU and others) in the moratorium area for the period 2013–2015 (Based on Presidential Instruction 6 of 2013).

Secondary bureaucracies

The same competence of secondary bureaucracies on title forest is also applied on forest area categories. 12 Ministries/Agencies sign Memoranda of Understanding on the Acceleration of forest area Stipulation in Indonesia which is witnessed by the President. The President also asks UKP-PP and Corruption Commission Eradication (KPK) to monitor and supervise this agreement (NFC, 2013b).

One of the root problems in forest area administration is the sheer number and discrepancy between existing maps. Several public agencies release data on the quantity and the quality of forest area. Therefore, UKP-PPP under the President instruction is responsible to initiate the creation by many related agencies and actors of a "One Map Policy", which captures degraded land and reconcile a large number of maps that can be used in the moratorium area (Satgas REDD+, 2013b). However, the Geospatial Information Agency (GIA) is "formally" appointed to provide Indonesian Geospatial Reference 2013 to implement the 'One Map Policy' (Forestry Research and Development Agency, 2013) and it is a formal responsibility attached on the Law No. 4 of 2011. Indonesian forest area boundary determination is not complete on the legal authorised forest map, resulting in conflict between MoFor and other stakeholders (Satgas REDD+, 2013b).

The discussion in Section 'Released forest area and change forest category' explains how rainforest transformation options could release forest area for non-forestry (and forestry) purposes or transform forest area into title forest. Secondary bureaucracies are also working on the process of releasing forest area to title forest use or state land for development. This has implications for SMNDP, which is responsible for the national space arrangement plan and by default is an agent in development uses of forest area (based on the Law No. 26 of 2007). UKP-PPP and Management Unit on REDD+ as presidential task unit for any special task on forest issues as well as NCCC (National Council on Climate Change) is also responsible for responding international forest regime.

The Ministry of Environment (MoE) is responsible for forest "environment", namely biodiversity. Indonesia ratified the United Nations Convention of Biodiversity (CBD) under Law No. 5 of 1994. Law No. 23 of 1997 on the Management of Environment established MoE as the core bureaucracy in charge of biodiversity. MoE is appointed as National Focal Point on CBD. MoE works closely with MoFor on forest area as the main steward of biodiversity issues.

Different with other secondary bureaucracies, the National Forestry Council (NFC) is not formed by government but it is formed by forestry multi-stakeholders organisation through Indonesian Forestry Congress IV on 2006 which comes from the representatives of non-government organisation, local community, forestry enterprise, regional governments, forestry professional association, and MoFor. This organisation as the interpretation of Forestry Law 41 of 1999 that government need stakeholders forum on forestry development. But on the other side, the legal letter of this organisation's structure is formed by MoFor which is indicated that NFC is a formal partner of MoFor. NFC as a multi-stakeholders forestry forum has three functions, including (1) government partners in making policy, (2) aligning communication between the parties, and (3) monitor the performance of forestry development (NFC, 2013a). The other multi-stakeholder based organisation which is legally partner with MoFor is Working Group on Tenure, and Working Group on Community Empowerment.

Legal options for tropical rainforests transformations

Jurisprudential options and bureaucratic responsibilities are explored below on the legal probabilities of the tropical rainforest transformation process. Options for transforming native forest into other land uses are described in Fig. 2. Steering to native forest transformation is discussed in Section 'Developing plantation and agroforestry outside forest area (title forest).'

Fig. 2 shows that basically, each of forest categories have possibilities to be transformed into all other categories. Logging and forest commodities are an essential and limiting factor in determining native forest transformation options in state forest area (specifically on productive and protection forest). Logging can only be conducted in title forests and production forests (or forest area relinquished by the state). Only forest commodities listed by MoFor can be planted in forest area. Consequently, it is almost impossible to set up palm oil plantations in production forests since forestry regulations do not categorise palm oil as a forest commodity. Exceptions exist if MoFor, MoA or regional government follow regulations on utilising forest area for non-forestry purposes, so all rainforest models (and commodities) are applicable to forest area. In spite of this, no forest area has recently been transformed to palm oil plantations. Commodity-based interest becomes an essential factor in transforming native forests into other uses. Forest moratorium policy and the MoA Regulation 98 of 2013 which limits only 100,000 hectares allowed to be developed as palm oil plantations have reduced the growth of palm oil transformation (The Indonesian Palm Oil Association, 2013; Investor Daily, 2013).

In addition to seizing control over forest area, central government prevents transformation of conservation forest into rubber or palm oil transformation models. One exception to this is if

Table 2
Main bureaucracies' responsible for forest area.

	MoFor	Regional Government (Province or District)
Specific Responsibilities on Conservation Forests	Administers and regulates duties of Managing Forest Conservation Unit (based on CNREL &FL)	Limited management responsibilities. Regional Government only manages Forest Park (*Tahura*). Based on Article 38 of Conservation, and Natural Resources and Its EcosystemLaw and Article 12 of Government Regulation 28 of 2011
Specific Responsibilities on Production Forests	Administers and regulates (FL)	Manages production forest by facilitating Forest Management Unit or issues rights on managing the forest area. Based on Forestry Law, Article 66 of Conservation and Natural Resources and Its Ecosystem Law, Article 7 of Government Regulation 38 of 2007, andArticle 8 of Government Regulation 6 of 2007
Specific Responsibilities on Protection Forests	Administers and regulates (FL)	Manages protection forest by facilitating Forest Management Unit or issuing rights on managing the forest area Based on Forestry Law, Article 66 of Conservation and Natural Resources and Its Ecosystem Law, Article 7 of Government Regulation 38 of 2007, and Article 8 of Government Regulation 6 of 2007

CNREL: Conservation and Natural Resources and Its Ecosystem Law No. 5 of 1990 also referred to as *UUK (Undang-undang konservasi SDA dan ekosistemnya)*.
FL: Forestry Law number 41 of 1999.

bureaucracies steer conservation forest area by transforming it into protection and production forest (see Section 'Released forest area and change forest category.')

Released forest area and change forest category

Released forest area. Following Forestry Law and Spatial Planning law (Law No. 26 of 2007), recent regulations on released forest area were formed on 2010 and 2012. Government Regulation 10 of 2010 on Forest Determination and Function Change Procedure (and Government Regulation No. 60 of 2012 on the amendment several points of Government Regulation 10 of 2010) offers two opportunities to change classification of forest, the first is partial alteration of forest area; or secondly, through province-wide proposal (Article 6). Forest cover of at least 30% of the total province or watershed must be maintained. The following points are some of the protocols for releasing forest area.

a Partial alteration of forest area; by way of (i) exchange of forest area (i.e. existing forest area become non-forest area in return for non-forest area becoming new forest area upon completion of the exchange), or (ii) relinquishment of forest area (i.e., existing forest area becomes non-forest area upon completion of the relinquishment).
b Exchange of forest area (land swap) can only be carried out in respect to: Permanent Production Forest, and/or, Limited Production Forest (Article 10). The exchange of forest areas can only be carried out on the basis of certain conditions and requirements (Article 11). This scheme shall be granted for a maximum period of 1 year from the issuance date, subject to the possibility of two years (Article 13).
c Relinquishment of forest area can only be carried out in respect to Convertible Production Forests (Article 19).

Revision of regional spatial planning as shown on Table 3 could be done each 5 years that Regional government (which is led by Province Government) is the key stakeholders based on the Spatial Planning law. The recent regulation (MoFor Regulation P.62 of 2013) has included the possibility to apply this legal option to

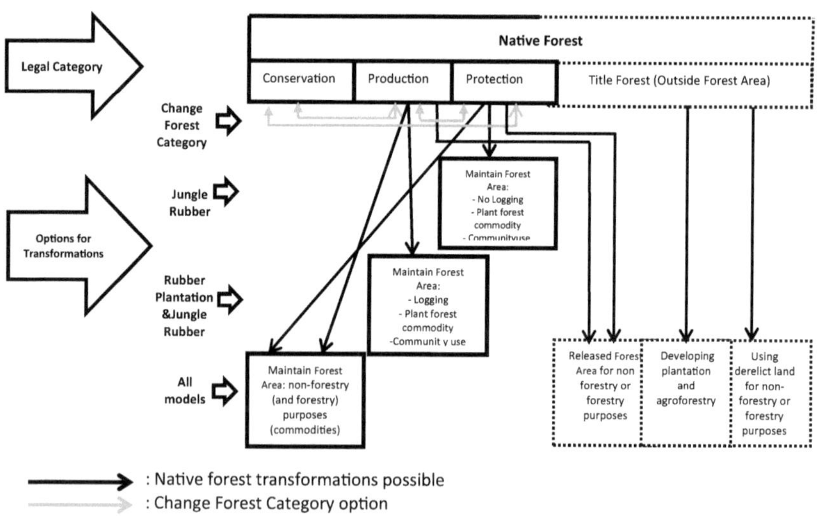

Fig. 2. Native forest transformation options.

able 3
ureaucracies releasing forest area for plantation.

	MoFor	Regional Government (District and or Province Government)
I. →Partial alteration (change) of forest area a. Exchange of forest area b. Relinquishment of forest area	→MoFor proposes forest area Relinquishment →Forms Integrated Team from scientific authority (to evaluate the feasibility of forest area release proposal) →Asks parliamentary opinion of the proposal →Issues the principal permit →Recommends land replacement →Guarantees province or watershed areas have a total forest area at least 30% of the total province or watershed area	→Proposes forest area →Relinquishment →Recommendation from governor or head of district
II. Province-wide proposal	→Forms Integrated Team from scientific authority (to evaluate the feasibility of forest area release proposal) →Asks parliamentary opinion of the proposal →Issuing Agreement	→Proposes revision of Regional Spatial Planning (Including Relinquishment forest area)

ource: Government Regulation No. 10 of 2010 on Forest Determination and Function Change Procedure and Law No. 26 of 2007 on Spatial Planning.

elease forest area for customary forest as explained in Section Recent legal developments on customary forest.'

Change forest category. Basically, Fig. 2 has showed us that all forest ategories have possibilities to be changed. Conservation area is the nost restricted area to release, however, it is possible to transform t into production and protection forest (then it is possible to be eleased). Government Regulation No. 10 of 2010 stated that forest rea alteration requires an application from MoFor or a proposal rom a governor of the relevant province. Furthermore, the alteration of forest area status to Convertible Production Forest Area annot be carried out in a province where the total forest area is ess than 30% of its total area.

Forest area for logging and forest plantation. In Article 28, paragraph 1) of FL, Forest Utilisation, including forest plantation or monoculture planting directly related to palm oil plantation model, is preferably carried out in **unproductive production forest** in order o preserve natural forests. Table 4 shows three opportunities for ogging and plantation in forest area.

MoFor has authority to issue licences for logging, indicating that entral government has power and control over the forest. One notable exception is the issuance of logging licences by regional government for the people forest plantation scheme.

Both natural forest concessions and industrial plantation forest schemes are regulated under MoFor Regulation P.50 of 2010. People forest plantation, community forest, and village forest schemes are explained in Section 'Recent legal developments on customary forest.' Natural forest concession lasts for 55 years and aims to utilise production forests, while Industrial plantation forest aims to develop production forest on plantation forest. This type of development is led by industrial groups to enhance the potential and quality of the production forest with the target of meeting the demands for basic industrial materials (lasting 100 years). People forest plantation is defined as forest built by local communities to enhance the potential and quality of production forest by implementing silviculture to ensure sustainability of forest resources lasting 60 years with an option for a single extension of 35 years).

In addition to estimating area and utilisation of Forest Plantation, the natural forest concession scheme amounted to about 23,647,239 ha, while the industrial plantation forest amounted to about 10,046,839 ha. The people forest plantation scheme amounted to about 165,410.06 ha over the same period (MoFor, 2012a).

Forest area for community forestry

Community forest which located on (state) forest area could be a choice for native forest transformation. These schemes is not confirming the community ownership towards forest area but strengthening their activities on forest management. It also maintains the relationship between MoFor as the land authority, regional government (both provincial and district government) as the bridge of bureaucracy between the national level and community, and local community or farmer groups as rights holders. These schemes will be effective in resolving existing conflicts and abuses (Marwa et al., 2010; Galudra et al., 2013; Nurrochmat, 2004; Nurrochmat et al., 2014a; similiar in village forest case on Akiefnawati et al., 2010; Supratman & Sahide, 2013). These relationships and built-in responsibilities appear in Table 5.

There are at least six schemes categorised as community forestry which are offered by the government (MoFor):

a. Village forest: the scheme built for the village welfare and managed by village institutions. It may be located on both production and protection forest, under MoFor Regulation P.49 of 2008 and MoFor Regulation 53 of 2011. MoFor has approved 83,401 ha for this scheme (MoFor, 2012b).
b. Community forest: the scheme built for community empowerment. It may be located on both protection and production forest, under MoFor Regulation P.37/Menhut-II/2007. It is managed by individual farmers or farmer groups. MoFor has approved 186,941 ha for this scheme (MoFor, 2012b).
c. People forest plantation: the scheme built for plantation forest development on production forest. Land is developed by an industrial group to enhance the potential and quality of the production forest in the frame of meeting the demands for basic industrial material production, under MoFor Regulation P.23 of 2007. MoFor has approved 700,831 ha for this scheme (MoFor, 2012b).
d. Partnership: collaboration between farmer groups and private companies (as forest utilisation permit holders) and Government (both MoFor and Regional Government) as the facilitator. This is the recent pattern of community forest regulated under MoFor Regulation P. 39 of 2013 on Community Empowerment through Forestry Partnership
e. Forest Area for Special Purpose: There are two main legal interpretation of 'special purpose'; the first is for forestry research, and the second is for respecting or recognising indigenous land rights managed by local communities. This regime is regulated under Government Regulation 6 of 2007. However, only 'special purpose' management units for research institutions have been developed in the field. No 'special purpose' management unit exists for local community stakeholders in the field (MoFor, 2012b).
f. Traditional Zone and Special Zone of National Park: one of the zoning rules on national park land that provides for special purposes including community forestry activities under MoFor Regulation P.56 of 2006.

Table 4
Schemes of Forest Timber Products Utilisation Business Licenses (IUPHHK) and bureaucratic responsibilities.

	MoFor	Regional Government
1. Natural Forests (Concession)(Government Regulation 3 of 2008)	Issuing licence for logging	Recommendation for logging permit.
2. Industrial Plantation Forests	Issuing licence for logging	Recommendation for logging permit.
3. People Forest Plantation	Issuing licence for logging	Issuing licence for logging
4. Village forest	Issuing licence for logging	Recommendation for logging permit.
5. Community forest	Issuing licence for logging	Recommendation for logging permit.

Table 5
Bureaucracies responsible for CFs schemes.

	MoFor	Regional Government
Village forests	Grant Licence for Logging (on production forest)	Issue recommendation and forest management right.
Community Forests	Grant Licence for Logging (on production forest)	Issue community forestry business right.
Forest Area for Special Purpose	Issue Management Right	Issue recommendations.
Partnership	Regulate and Facilitate stakeholder interactions	Facilitate farmers' groups and business or permit holders.
People Forest Plantation	Grant Licence for Logging (on production forest)	Issue recommendation and grant licence for logging (on production forest).
Traditional zone and Special Zone of National Park (Conservation area)	The head of conservation unit (National Park) on behalf of MoFor Issuing Management Rights	Issue management rights.

MoFor and Regional Government are the key stakeholders on regulating and managing these community forestry schemes. Table 5 provides a summary of the bureaucracies involved and their responsibilities.

Forest area for community use is somehow interpreted as developing agroforestry systems (including non-timber forest products and ecosystem services). Even though MoFor Regulation P.35 of 2007 has listed non timber forest products (NTFs) as their commodities, but MoA as the core bureaucracy on developing agroforestry products (including NTFPs) find difficulties in developing agroforestry products which are grown on forest areas due to the restriction access of forest area.

Using forest area for non-forestry purposes and the palm oil commodity policy discourse

The previous policy on MoFor Regulation P. 62 of 2011 allowed palm oil to be planted on production forest under the guidelines of the industrial forestry concession scheme. After community dialogues raised public discourse on the policy, MoFor nullified Regulation P.64 of 2011. There is, nevertheless, the possibility for palm oil to be planted in forest area. Under FL Article 38, this could be performed by proposing a scheme to use production and protection forests for non-forestry purposes in the interest of national development. If the actors agree or meet political consensus that palm oil and rubber plantations are categorised as "certain agricultural and food security in the context of energy security", under Article 1 of Government Regulation Number 61 of 2012, then palm oil and rubber commodities could be grown in forest area. Furthermore, this regulation provides the use of forest area for non-forestry purposes based on forest use permits. In 2010, seventy-seven plots covering about 43,171.96 ha of forest area were approved for mining and non-mining activities, while sixty-eight plots covering area of about 60,313.47 ha were used for mining and non-mining activities under land compensation (MoFor, 2012a).

Developing plantation and agroforestry outside forest area (title forest)

This option is identical to the legal entity-forest category of title forests which actually is basically from a set of laws and regulations and has been explained in our category of title forest in Section 'Bureaucracies responsible for title forest.' This legal entity forest could be developed from the right of cultivation, or business permit (*HGU*). This option is possible to conduct any plantation and agroforestry activities in land outside forest area.

Using waste land (non forest area)

Government Regulation 11 of 2010 on Controlling and Using Waste Land has been issued to maximise waste land that NLA has indicated 4.8 million ha waste land in Indonesia (Mongabay Indonesia, 2012). Waste land is all land that is not maintained by the owner and inhabitants for all land permits such as HGU (plantation business land), state land, Non Forested Area (APL) or other title lands. This regulation asks NLA as core bureaucracies on implementing this regulation through coordination with MoA and MoFor which are the most related stakeholders. MoFor argued that this regulation is a good opportunity to use waste land for palm oil plantations (State-Owned Enterprises, 2011).

Steering back to Native Forest

Transformation back to native forest can be done through legal forest restoration by business and through national programmes for land rehabilitation and reforestation.

a. **Ecosystem Restoration scheme**: the legal basis for this scheme is MoFor Regulation P.50 of 2010, and the purpose of this scheme is to develop a zone within a production forest site to preserve and enhance the quality of its ecosystem. These activities include maintenance, protection and restoration of the forest ecosystem, including planting, enriching, spacing, breeding of fauna, returning flora and fauna to wildlife, and non-biological elements (soil, climate and topography) to endemic species, so as to achieve equilibrium in bio-data and the ecosystem. Hence, all models of native forest transformation located in production forest are eligible to access this scheme. Roughly 525,340 hectares have been approved for this scheme (MoFor, 2012a).
b. **Forest and Land Rehabilitation:** this steering applies in both forest area and title forest and funded by MoFor. Forest management units under regional government or MoFor authority are eligible to propose rehabilitation using their own budgetary resources. Originally this activity was based on the Joint Decrees of the Coordinating Ministry of Welfare, Coordinating Ministry of the Economy and Coordinating Ministry of Politics and Security Numbers 09 of 2003, 16of 2003,08 of 2003 on the Establishment Coordination Team Improvements through Environmental Rehabilitation and National Afforestation. The following year MoFor issued ministerial decisions for implementation, namely Decision 340 of 2003 and Regulation Number 25 of 2006. Regional government is also authorised to propose Forest and Land Rehabilitation in title forest. Hence, all models of rainforest transformation are eligible to access this scheme.

During the period 2006–2010 the Ministry of Forestry has managed the reforestation of 149,422.75 ha under this programme (MoFor, 2012a).

Conclusion

Bureaucratic responsibilities influence tropical rainforest transformarmations systems

This paper finds that there are six legal forest land use categories in the dual land system of Indonesian territory. There are three categories of forest area, which include (1) conservation forest, (2) production forest (3) protection forest; and three categories of title forest, namely (1) private forest (private individual ownership), (2) plantation forest (legal entity ownership), and (3) customary forest. 'Legality entity-plantation forest' is categorised on title forest due to MoFor or other bureaucracies still regulate logging mechanism of trees production from this plantation, even though some international definitions do not categorise plantation forests as 'forest'. Four types of rainforest transformation (native forest, jungle rubber or agroforestry, rubber plantation, and palm oil plantation) are present in this study across all legal forest land use categories. The core bureaucracies steering tropical rainforest transformation are Ministry of Forestry, National Land Agency, regional government (which relies mostly on district government as an autonomous region entity). An intersection of responsibilities exist between a number of these agencies, but it is most significant between Ministry of Forestry, regional government and National Land Agency due to the right of licensing power these bureaucracies share (such as; allocating regional land use, releasing forest area, rights of cultivation, rights of logging). Commodity and product based interests also create cross intersections of responsibility between the Ministry of Forestry and Ministry of Agriculture in 'agroforestry' commodities. Legal land status and Commodity interests have become an essential consideration in the transformation of native forest to other types of land use. This is in line with bureaucratic politic theory that they are competing for resources and political domain (Allison, 1971; Niskanen, 1971; Stern, 1998; Krott, 1990; Hubo & Krott, 2010; Peters, 2010; Giessen, 2011).

The secondary bureaucracies steering native forest transformation are State Ministry for National Development Planning/National Development Planning, the Ministry of Trade, the Ministry of Industry, the Ministry of Energy and Mineral Resources the Ministry of Environment, the Ministry of Manpower and Transmigration, Presidential Task Force Unit for Supervision and Management of Development, the Ministry of the Environment Geospatial Information Agency, the National Commission on Human Rights, National Council on Climate Change, Management unit of REDD+, and the National Forestry Council. The President has appointed the Presidential Task Force Unit for Supervision and Management of Development and asks Corruption Eradication Commission to monitor the memorandum of understanding among 12 ministries/agencies on forestry land tenure governance. This indicates that the President using his coercive power through a super bureaucracy to lead beyond the 'traditional agencies' to accelerate special mission. It is in line with Krott (2005) that the main sources of bureaucracies' power based on coercion are their legal mandates (including decision rights and sanction mechanisms), technical resources (enabling bureaucracies to maintain a policy process).

The existence of National Council on Climate Change, Management unit of REDD+, and the Presidential task unit which is directly involved on the REDD+ regime in Indonesia indicate that the international forest regime seems to be relevant in many bureaucracies in Indonesia which is in line with Wibowo & Giessen (2012) and Giessen (2013a, 2013b).

Legal options for tropical rainforest transformations

This paper finds a number of legal options for steering native forest transformation tied to formal bureaucratic responsibilities. Native forest transformation options for forest area include a release of forest area for any purposes, using forest area for non-forestry purposes (e.g. for food and energy security), maximising production forest for logging, and developing community forest schemes. In title forest, transformation options are plantation forest, developing plantation and agroforestry, and taking full advantage of wasteland. Other option is steering back to native forest applied in both forest area and title forest. Native forest transformation options reflect the tendency to take high pressure on forest area with the most pressure coming from releasing forest area options and using forest area for non-forestry purposes. Customary forest creates additional constraints on the quantity of forest area. However, transformation back to native forest can be done through legal forest restoration by business and through national programmes for land rehabilitation and reforestation.

Hypotheses on informal bureaucratic interests for future research

The informal responsibilities and interest-based motives described in bureaucratic theory demonstrate that bureaucracies pursue formal as well as informal goals (Krott, 1990, 2005; Peters, 2010). Based on that theory the authors analyse the outcomes of formal tasks to conceptualise a hypothesis on the interests of bureaucracies for further research. The central hypotheses contends that both core and secondary bureaucracies have conflicting interests over all four types of tropical rainforest transformation due to the areas of overlapping responsibility. A second claim is that MoFor acts to maintain forest area cover quantity in a manner that is in line with bureaucratic competition for resources and responsibility in policy domains (Allison, 1971; Niskanen, 1971; Stern, 1998; Krott, 1990; Hubo & Krott, 2010; Peters, 2010). These will also contribute to the 'tough fight' of political forest definition on the law level as well as the implementation of all forest types. For example, in the case of Constitutional Court decision 35 of 2012 customary forest should be transformed from forest area to title forest, then Ministry of Forestry will behave to keep and provide regulation on the strict definition and requirement of customary forest as the part of efforts to maintain the quantity of forest area cover. Ministry of Forestry is using (and 'optimising') community forest schemes to campaign for indigenous rights' interests over those of customary forest. In other cases on finalising forest area boundaries or mediating Ministry of Forestry and other core and secondary bureaucracies on the conflict of forest boundary between forest area and other land use sector, Ministry of Forestry will take forest area cover quantity as the main attention on the mediation between Ministry of Forestry and other sector bureaucracies such as National Land Agency, Ministry of Agriculture, and regional government (similar case on Giessen & Krott, 2009; Krott & Hasanagas, 2006; Rayner et al., 2001). Third, regional government competes with Ministry of Forestry over native forest transformation to palm oil and rubber plantation, due to local income generation from both crops. Regional governments typically choose palm oil plantations over rubber agroforestry and plantations as well as REDD initiative due to the higher income potential from palm oil commodities. This is in line with van Noordwijk et al. (2013) that national interest in retaining global palm oil exports gained priority over expectations of REDD forest rents. But this competition will be camouflaged by keeping conservation issues invisible, neglecting them within administrative procedures (similar in Lower Saxony case by Hubo & Krott, 2013). Fourth, on the agroforestry development, Ministry of Forestry and Ministry of Agriculture will compete for the domain responsibilities due to the commodity based policy and the impact of dual land systems. This contestation is amplified by the nature of bureaucracies which is divided into production oriented bureaucracies and conservation oriented bureaucracies (Hirsch & Warren, 1998; Giessen et al., 2014). For example, even though National Park Management has been mandated on approving special zone and traditional zone for community use (agroforestry) but it will be hard to implement in the field due to the contradiction of their basic position and interest as the conservation oriented bureaucracies. Fifth, Palm oil as the very high value commodity for national interest (SMNDP, 2011; Caroko et al., 2011) will enhance the influence of related production oriented bureaucracies and face high challenges from conservation oriented bureaucracies. Sixth, secondary bureaucracies such as Presidential Task Force Unit for Supervision and Management of Development, National Council on Climate Change, the National Forestry Council, Ministry of Environment, State Ministry for National Development Planning/National Development Planning, and REDD+ management unit, possesses much influence in international forest issues, and therefore these bureaucracies find abundant financial support from international, regional, and bilateral donors. This can potentially lead to struggles between these secondary bureaucracies and the core bureaucracies on finalising forest demarcation, registering customary forest, and granting rights of plantation, all due to competition in budget sharing. Seventh, strong presidential representation in delegations to the Presidential Task Force Unit for Supervision and Management of Development, National Council on Climate Change, and REDD+ management unit are found in this research to be an illustration of presidential coercive power. Eighth, strong environmentalist and community based mission from the influence of international forest regimes will amplify the contestation between conservation oriented bureaucracies and production oriented bureaucracies in national and sub national levels. The influence of international forest regimes is felt in the strength of actors implementing programmes and overseeing supporting budgets in both domestic and international donor budgets (Bernstein & Cashore, 2012). Whether bureaucratic competition and contestation is reduced or amplified is dependent on beneficiary responsibility sharing in mandatory delivery power, and achievement of resources. Ninth, key staff and leaders of presidential task force bureaucracies come from academic and third sector backgrounds. This affects the ways these organisations are managed and their institutional missions, at times contradicting organisational thinking with traditional bureaucracy like Ministry of Forestry. This is in line with bureaucratic politics' assumption that parochial, personal, national, and domestic interests shape the actors' positions towards the issue at hand (Tagma & Uzun, 2012).

Acknowledgements

We greatly appreciate the financial support from DFG research project on the fragmentation of the international forest regime complex (PAK 813), the Eva Mayr-Stihl Foundation as well as the Indonesian Government Fund (DIKTI scholarship).

References

Allison, G.T., 1971. Essence of Decision: Explaining the Cuban Missile Crisis. Little, Brown and Company, Boston, pp. 1971.

Akiefnawati, R., Villamor, G.B., Zulfikar, F., Budisetiawan, I., Mulyoutami, E., Ayat, A., Noordwijk, M.V., 2010. Stewardship agreement to Reduce Emissions from Deforestation and Degradation (REDD): case study from Lubuk Beringin's Hutan Desa, Jambi Province, Sumatra, Indonesia. Int. For. Rev. 12 (4), 349–360, http://www.bioone.org/doi/full/10.1505/ifor.12.4.349

Anaya, S.J., Grossman, C., 2002. Case of Awas Tingni v. Nicaragua: a step in the International Law of Indigenous Peoples. Ariz. J. Int'l Comp. L 19, 1.

Anderson, J., 2000. Four considerations for decentralized forest management: subsidiarity, empowerment, pluralism and social capital. In: Enters, T., Durst, P.B., Victor, M. (Eds.), Decentralization and Devolution of Forest Management in Asia and the Pacific. , pp. 17–27.

Andiko, 2010. Legal analysis protection of high conservation value areas in palm plantations in Indonesia. HCV Resour. Netw.

AMAN, 2013. Nota Kesepahaman AMAN, Available at http://www.aman.or.id/mou/ (last accessed 01.10.2013).

Arizona, Y., 2013. Mahkamah Konstitusi dan Reformasi Tenurial Kehutanan, Available at https://www.academia.edu/5470629/Mahkamah_Konstitusi_dan_Reformasi_Tenurial_Kehutanan, (last accessed 10.01.2014).

Arizona, Y., Mary, S.R., Nagara, G., 2012. ANOTASI PUTUSAN MK NO.45/PUU-IX/2011 mengenai Pengujian Konstitusionalitas Kawasan Hutan dalam Pasal 1 Angka 3 UU No.41 tahun 1999 tentang Kehutanan. HuMa.

Bakker, L., Moniaga, S., 2010. The space between: land claims and the law in Indonesia. AJSS 38 (2), 187–203, http://dx.doi.org/10.1163/156853110X490890.

Bernstein, S., Cashore, B., 2012. Complex global governance and domestic policies: four pathways of influence. Int. Affor. Areairs 88 (3), 585–604.

Beukema, H., van Noordwijk, M., 2004. Terrestrial pteridophytes as indicators of a forest-like environment in rubber production systems in the lowlands of Jambi, Sumatra. Agric. Ecosyst. Environ. 104 (1), 63–73, http://dx.doi.org/10.1016/j.agee.2004.01.007.

Brockhaus, M., Obidzinski, K., Dermawan, A., Laumonier, Y., Luttrell, C., 2012. An overview of forest and land allocation policies in Indonesia: is the current framework sufficient to meet the needs of REDD+? For. Policy Econ. 18, 30–37, http://dx.doi.org/10.1016/j.forpol.2011.09.004.

Budiman, A., Penot, E.,1997. Smallholder rubber agroforestry in Indonesia. In: International Rubber Conference. IRC.

Caroko, W., Komarudin, H., Obidzinski, K., Gunarso, P.,2011. Policy and institutional frameworks for the development of palm oil-based biodiesel in Indonesia. In: Working Paper 62. CIFOR, Bogor, Indonesia.

Casson, A., 1999. The Hesitant Boom: Indonesia's Oil Palm Sub-Sector in an Era of Economic Crisis and Political Change. Center for International Forestry Research.

Comte, I., Colin, F., Grunberger, O., Follain, S., Whalen, J.K., Caliman, J.P., 2013. Landscape-scale assessment of soil response to long-term organic and mineral fertilizer application in an industrial oil palm plantation, Indonesia. Agric. Ecosyst. Environ. 169, 58–68, http://dx.doi.org/10.1016/j.agee.2013.02.010.

Fay, C., Sirait, M., ICRAF Southeast Asia Working Paper (No. 2005_3) 2005. Kerangka Hukum Negara dalam Mengatur Agraria dan Kehutanan Indonesia: Mempertanyakan Sistem Ganda Kewenangan atas Penguasaan Tanah.

Feintrenie, L., Schwarze, S., Levang, P., 2010. Are Local People Conservationists? Analysis of Transition Dynamics from Agroforests to Monoculture Plantations in Indonesia. Ecol. Soc. 15 (4), 37.

Fitzpatrick, D., 2006. Evolution and chaos in property rights systems: the third world tragedy of contested access. Yale Law J., 996–1048, http://dx.doi.org/10.2307/20455644.

Forestry Research and Development Agency, 2013. One Map Policy: Momen Kebangkitan Penelitian Kehutanan Berbasis Informasi Geospasial, Available at http://www.forda-mof.org/index.php/berita/post/1554 (last accessed 30.10.2013).

Friedrichs, J., 1990. MethodenempirischerSozialforschung. Westdeutscher Verlag, Opladen, pp. 429.

Galudra, G., van Noordwijk, M., Agung, P., Suyanto, S., Pradhan, U., 2013. Migrants, land markets and carbon emissions in Jambi, Indonesia: land tenure change and the prospect of emission reduction. Mitig. Adapt. Strateg. Global Change, 1–17, http://dx.doi.org/10.1007/s11027-013-9512-9.

Giessen, L., Krott, M., Möllmann, T., 2014. Increasing representation of states by utilitarian as compared to environmental bureaucracies in international forest and forest-environmental policy negotiations. For. Policy Econ. 38, 97–104, http://dx.doi.org/10.1016/j.forpol.2013.08.008.

Giessen, L., 2013a. Reviewing the main characteristics of the international forest regime complex and partial explanations for its fragmentation. Int. For. Rev. 15 (1), 60–70, http://dx.doi.org/10.1505/146554813805927192.

Giessen, L., 2013b. Fragmentierung als Schlüsselfaktor des internationalen Waldregimes:von einem mono- zu einem multi-disziplinären methodischen Rahmen für eine vertiefte Forstpolitikforschung. Allg. For. Jagdztg. 184 (3–4), 47–57.

Giessen, L., 2011. Reviewing empirical explanations of policy change: options for its analysis and future fields of research. Allg. For. Jagdztg. 182 (11–12), 248–259.

Giessen, L., Krott, M., 2009. Forestry joining integrated programmes? A question of willingness, ability and opportunities. Allg. For. Jagdztg. 180 (5–6), 94–100.

Hirsch, P., Warren, C., 1998. Introduction. Through the Environmental Looking Glass. The Politics of Resources and Resistance in Southeast Asia. The Politics of the Environment in Southeast Asia. Resources and Resistance. Routledge, Houndmills, pp. 1–28.

Hubo, C., Krott, M., 2013. Conflict camouflaging in public administration—a case study in nature conservation policy in Lower Saxony. For. Policy Econ. 33, 63–70, http://dx.doi.org/10.1016/j.forpol.2012.10.008.

Hubo, C., Krott, M., 2010. Politiksektoren als Determinante von Umweltkonflikten am Beispiel invasiver gebietsfremder Arten. In: Feindt, P.H., Saretzki, T. (Hrg.)(Eds.), Umwelt- und Technikkonflikte. VS Verlag für Sozialwissenschaften, Wiesbaden.

Hunt, C.O., Rabett, R.J., 2013. Holocene landscape intervention and plant food production strategies in island and mainland Southeast Asia. J. Arch. Sci., http://dx.doi.org/10.1016/j.jas.2013.12.011.

HuMa, 2013. The Customary Forests after the Constitutional Courts Ruling, Available at http://www.inaturefilms.org/the-customary-forests-after-the-constitutional-courts-ruling/ (last accessed 01.10.2013).

Indonesian Constitutional Court, 2011. Risalah Sidang Perkara Nomor 45/PUU-IX/2011.

Investor Daily, 2013. Pengusaha Sawit Tolak Pembatasan Luas Kebun, Available at http://www.investor.co.id/agribusiness/pengusaha-sawit-tolak-pembatasan-luas-kebun/71220 (last accessed 07.11.2013).

JKPP, 2013. Pemetaan Partisipatif: Komponen Dasar Pengakuan dan Pemetaan Wilaah Adat, Available at http://www.unorcid.org/index.php/about-redd/redd-in-the-news/redd-indonesia/180-national-workshop-on-the-forestry-law-constitutional-court-ruling-no-35-puu-x-2012-speeches-and-presentations-are-now-available (last accessed 29.10.2013).

Krott, M., 1990. Öffentliche Verwaltung im Umweltschutz. Ergebnisse einer behördenorientierten Policy-Analyse am Beispiel Waldschutz. W. Braumüller Verlag, Wien.

Krott, M., 2005. Forest Policy Analysis. Springer, Dordrecht, pp. 323pp.

Krott, M., Hasanagas, N.D., 2006. Measuring bridges between sectors: causative evaluation of cross-sectorality. For. Policy Econ. 8 (5), 555–563, http://dx.doi.org/10.1016/j.forpol.2005.07.004.

Larson, A.M., Barry, D., Dahal, G.R., 2010. Tenure change in the global south. In: Larson, A.M., Barry, D., Dahal, G.R., Colfer, C.J.P. (Eds.), Forests for People: Community Rights and Forest Tenure Reform. Earthscan, pp. 3–18.

Marwa, J., Purnomo, H., Nurrochmat, D.R., 2010. Managing the Last Frontier of Indonesian Forest in Papua. IPB-AKECOP Korea, Bogor.

Michon, G., de Foresta, H., 1995. The Indonesian agroforest model forest resource management and biodiversity conservation. In: Halladay, P., Gilmour, D.A. (Eds.), Conserving Biodiversity Outside Protected Areas: The Role of Traditional Agro-Ecosystems. IUFRO, pp. 90–106.

Ministry of Agriculture, 2013. Production, Area and Productivity Estate Crops in Indonesia. In: Produksi, Luas Areal dan Produktivitas Perkebunan di Indonesia.

Ministry of Forestry, 2012a. Forestry Statistic of Indonesia., pp. 2011.

Ministry of Forestry, 2012b. Data dan Informasi Ditjen Forestry Planning Kehutanan Tahun 2012.

Ministry of Forestry, 2005. Hutan Rakyat Indonesia Sangat Prospektif untuk Industri Kehutanan.

Ministry of Energy and Mineral Resources, Available at http://www.esdm.go.id/berita/43-mineral/829-pengembangan-biofuel-perlu-insentif.html (last accessed 01.12.2013) 2013. Pengembangan Biofuel Perlu Insentif.

Mongabay Indonesia, 2012. http://www.mongabay.co.id/2012/12/28/konflik-agraria-2012-ratusan-petani-ditahan-dan-25-tertembak/ (last accessed 01.12.2013).

Movuh, Y., Mbolo, C., 2012. The Colonial heritage and post-Colonial influence, entanglements and implications of the concept of community forestry by the example of Cameroon. For. Policy Econ. 15, 70–77, http://dx.doi.org/10.1016/j.forpol.2011.05.004.

NFC, 2013a. Sejarah Dewan Kehutanan Nasional, Available at http://dkn.or.id/tentang-kami/sejarah/ (last accessed 06.11.2013).

NFC, 2013b. Policy Brief of National Forestry Council (NFC) on the Memorandum of Understanding of 12 Ministries and Institutions on the Acceleration of Forest Area Stipulation in Indonesia. Forest Area Restructuring For the Revival of National Forestry. Issue, Policy and Role of National Forestry Council.

Niskanen, W., 1971. Bureaucracy and Representative Government. Aldine - Atherton, Chicago.

Nurrochmat, D.R., Hasan, M.F., Suharjito, D., Hadianto, A., Ekayani, M., Purwawangsa, H., Ryandi, E.D., 2012. Ekonomi politik kehutanan: Mengurai mitos dan fakta pengelolaan hutan. INDEF, Jakarta.

Nurrochmat, D.R., 2004. Social forestry to alleviate poverty: concepts and experiences in Indonesia. In: Hebel, J., Samadi, D., Nurrochmat (Eds.), Poverty Alleviation: Concepts and Experiences: Focused on Indonesian Cases. Cuvillier Verlag, Göttingen, pp. 39–44.

Nurrochmat, D.R., 2005. The Impacts of Regional Autonomy on Political Dynamics, Socio-Economics, and Forest Degradation. Cuvillier Verlag, Goettingen.

Nurrochmat, D.R., Darusman, D., Ruchjadi, D., 2014a. Rekonstruksi sistem tenurial kehutanan (reconstruction of forest tenure system). Risal. Kebijak. Kehutan. Vol. 1, 2014, No 1 April.

Nurrochmat, D.R., Dharmawan, A.H., Obidzinski, K., Dermawan, A., Erbaugh, J.T., 2014b. Contesting national and international forest regimes: Case of timber legality certification for community forests in Central Java, Indonesia. J. For. Policy Econ., http://dx.doi.org/10.1016/j.forpol.2014.09.008 (in press).

Partohardjono, S., Pasaribu, D., Fagi, A., 2005. The Forest Margins of Sumatra, Indonesia,., pp. 291–304, Available at http://www.asb.cgiar.org/PDFwebdocs/The%20Forest%20Margins%20of%20Sumatra%20Indonesia.pdf (last accessed on 05.09.2013).

Peluso, N.L., Vandergeest, P., 2001. Genealogies of the political forest and customary rights in Indonesia, Malaysia, and Thailand. J. Asian Stud. 60 (3), 761–812, 10.2307/2700109.

Peluso, N.L., 1995. Whose woods are these? Counter-mapping forest territories in Kalimantan, Indonesia. Antipode 27 (4), 383–406, http://dx.doi.org/10.1111/j.1467-8330.1995.tb00286.x.

Peluso, N.L., 1992. Rich Forests, Poor People: Resource Control and Resistance in Java. Univ of California Press.

Peters, B.G., 2010. The Politics of Bureaucracy – An Introduction to Comparative Public Administration, 6th ed. Routledge, Oxon.

Rayner, J., Howlett, M., Wilson, J., Cashore, B., Hoberg, G., 2001. Privileging the sub-sector: critical sub-sectors and sectoral relationships in forest policy-making. For. Policy Econ. 2 (3), 319–332, http://dx.doi.org/10.1016/S1389-9341(01)00038-7.

Satgas REDD+, 2013a. Mengakui Hutan Adat dan Masyarakat Adat, Available at http://www.satgasreddplus.org/component/k2/item/99-mengakui-hutan-adat-dan-masyarakat-adat (last accessed 29.10.2013).

Satgas REDD+, 2013b. 1 Tahun Pelaksanaan Inpres 10/2011: Penundaan Pemberian Izin Baru dan Penyempurnaan Tata Kelola pada Hutan Alam Primer dan Lahan Gambut.

Simon, H.A., 1981. Entscheidungsverhalten in Organisation. Deutsche Übersetzung 3. Landsberg, Auflage.

Sirait, M., Fay, C., & Kusworo, A., 2001. Bagaimana hak-hak masyarakat adat dalam mengelola sumber daya alam diatur. Seri Kebijakan I, Maret 2011, pp 1-38. Published by ICRAF, LATIN, & P3AE-UI. Available at http://www.worldagroforestry.org/sea/Publications/files/book/BK0047-04.PDF.(last accessed 07.11.13).

Smith, J., Obidzinski, K., Subarudi, S., Suramenggala, I., 2003. Illegal logging, collusive corruption and fragmented governments in Kalimantan, Indonesia. Int. For. Rev. 5 (3), 293–302.

SMNDP, 2011. The Masterplan for Acceleration and Expansion of Indonesia Economic Development (MP3EI).

State-Owned Enterprises, 2011. Menhut: Lahan terlantar bisa dipakai untuk kebun sawit, Available at http://www.bumn.go.id/ptpn4/id/publikasi/indonesia-menhut-lahan-terlantar-bisa-dipakai-untuk-kebun-sawit/ (last accessed 07.11.2013).

Steffan-Dewenter, I., Kessler, M., Barkmann, J., Bos, M.M., Buchori, D., Erasmi, S., Forest areaust, H., Gerold, G., Glenk, K., Gardstein, S.R., Guhardja, E., Hartereld, M., Hertel, D., Höhn, P., Kappas, M., Köhler, S., Leuscher, C., Maertens, M., Marggraf, R., Migge-Kleian, S., Mogea, J., Pitopang, R., Schaefer, M., Schwarze, S., Sporn, S.G., Steingrebe, A., Tjitrosoedirdjo, S.S., Tjitrosoemito, S., Twele, A., Weber, R., Woltmann, L., Zeller, M., Tscharntke, T., 2007. Tradeoffs between income, biodiversity, and ecosystem functioning during tropical rainforest conversion and agroforestry intensification. Proc. Natl. Acad. Sci. 104 (12), 4973–4978.

Stern, E., 1998. Wither the study of governmental politics in foreign policymaking? A symposium. Mershon Int. Stud. Rev. 42, 205–255.

Sunderlin, W.D., Larson, A.M., Cronkleton, P., 2009. Forest tenure rights and REDD. Realising REDD. In: Angelsen, A., Brockhaus, M. (Eds.), Realising REDD+: National strategy and policy options. CIFOR, pp. 139–150.

Supratman, Sahide, M.A.K., 2013. Hutan Desa dan Pembangunan Sosial Ekonomi Masyarakat Desa di Kabupaten Bantaeng. Ministry of Forestry and Partnership.

Swaine, M.D., Whitmore, T.C., 1988. On the definition of ecological species groups in tropical rain forests. Vegetatio 75 (1–2), 81–86.

Tagma, H.M., Uzun, 2012. Bureaucrats, ayatollahs, and Persian politics: explaining the shift in Iranian nuclear policy. Korean J. Def. Anal. Vol. 24 (June (2)), 239–264.

The Indonesian Palm Oil Association, 2013. Available at http://www.gapki.or.id/Page/NewsDetail?guid=3e021776-87e4-42a1-b5b7-4fe535d56d2c (last accessed 07.11.13).

UKP-PPP, 2013. Peta Indikatif Penundaan Izin Baru, Available at http://www.ukp.go.id/peta-indikatif-penundaan-izin-baru (last accessed 29.10.2013).

van Noordwijk, M., Agus, F., Dewi, S., Purnomo, H., 2013. Reducing emissions from land use in Indonesia: motivation, policy instruments and expected funding streams. Mitig. Adapt. Strateg. Global Change, 1–16, http://dx.doi.org/10.1007/s11027-013-9502-y.

Walter, J.M.N., Torquebiau, E.F., 1997. The geometry of the canopy of a dipterocarp rain forest in Sumatra. Agric. For. Meteorol. 85 (1), 99–115.

Weber, M., 1922. Wirtschaft und Gesellschaft Grundriß der verstehenden Soziologie. Studienausgabe. Zitiert nach der Ausgabe von 1988, Tübingen.

Wells, P., Franklin, N., Gunarso, P., Paoli, G., Mafira, T., Kusumo, D., Clanchy, R.B., 2012. Indonesian Constitutional Court Ruling Number 45/PUU-IX/2011 in relation to Forest Lands. Implications for Forests, Development and REDD+. Final Draft Daemeter Consultant. TBI Indonesia. Makarim & Taira S Consultant.

Wibawa, G., Joshi, L., Van Noordwijk, M., Penot, E., 2006. Rubber based Agroforestry Systems (RAS) as Alternatives for Rubber Monoculture System. IRRDB Annual Conference, Ho-chi-minh city, Vietnam.

Wibowo, A., Giessen, L., 2012. Identifying International Forest-Related Issues in Indonesia based on actors' statements in public and expert deliberations. Int. J. Soc. For. 5 (1).

Wicke, B., Sikkema, R., Dornburg, V., Faaij, A., 2011. Exploring land use changes and the role of palm oil production in Indonesia and Malaysia. Land Use Policy 28 (1), 193–206, http://dx.doi.org/10.1016/j.landusepol.2010.06.001.

Wollenberg, E., Moeliono, M., Limberg, G., Iwan, R., Rhee, S., Sudana, M., 2006. Between state and society: local governance of forests in Malinau, Indonesia. For. Policy Econ. 8 (4), 421–433, http://dx.doi.org/10.1016/j.forpol.2005.08.012.

Zen, Z., Barlow, C., Gondowarsito, R.,2005. Oil palm in Indonesian socio-economic improvement: a review of options. In: Working/Technical Paper. Australian National University, Available at http://hdl.handle.net/1885/43005 (last accessed 21.10.2013).

Indonesian laws and legal regulation documents

Constitutional Court Decision 35 of 2012. *Keputusan Mahkamah Konstitusi Nomor 35/PUU-X/2012.*

Constitutional Court Decision 45 of 2011. *Keputusan Mahkamah Konstitusi Nomor 45/PUU-IX/2011.*

Government Regulation 6 of 2007 on Forest Governance and Forest Management Plan, and Utilization of Forest. *Peraturan*

Pemerintah Nomor 6 Tahun 2007 tentang Tata Hutan dan Penyusunan Rencana Pengelolaan Hutan, Serta Pemanfaatan Hutan.

Government Regulation 8 of 1953 on Mastery of State Land. *Peraturan Pemerintah Nomor 8 tahun 1953 tentang Penguasaan Tanah-Tanah Negara.*

Government Regulation 10 of 2010 on Forest Determination and Function Change *Procedure Peraturan Pemerintah Nomor 10 Tahun 2010 tentang Tata Cara Perubahan Peruntukan dan Fungsi Kawasan Hutan.*

Government Regulation 11 of 2010 on Controlling and Using Waste Land. *Peraturan Pemerintah Nomor 110 Tahun 2010 Tentang penertiban dan pendayagunaan tanah terlantar.*

Government Regulation 24 of 2010 on Using State forest area. *Peraturan Pemerintah Nomor 24 Tahun 2010 tentang Penggunaan Kawasan Hutan.*

Government Regulation No. 26 of 2008 on the National Spatial Plan. *Peraturan Pemerintah No. 26 Tahun 2008 tentang Rencana Tata Ruang Wilayah Nasional.*

Government Regulation 27 of 1999 on Environment Impact Asessment. *Peraturan Pemerintah No. 27 Tahun 1999 Tentang Analisis Mengenai Dampak Lingkungan Hidup.*

Government Regulation 28 of 2011 on the management of nature reserves and conservation areas. *Peraturan Pemerintah No. 28 Tahun 2011 tentang Pengelolaan kawasan suaka alam dan kawasan pelestarian alam.*

Government Regulation 38 of 2007 on Government Afforest areairs Division of the Government, Provincial Government and the District/Municipality Government. *Peraturan Pemerintah Nomor 38 Tahun 2007 tentang Pembagian Urusan Pemerintahan antara Pemerintah, Pemerintah Daerah Provinsi dan Pemerintah Daerah Kabupaten/Kota.*

Government Regulation 40 of 1996 on right of cultivation-business permit, right of use of structures, and right of use. *Peraturan Pemerintah Nomor 40 Tahun 1996 tentang Hak Guna Usaha, Hak Guna Bangunan dan Hak Pakai Atas Tanah.*

Government Regulation 40 of 1996 on Investment Plan For Indonesia. *Peraturan Pemerintah Nomor 40 Tahun 1996 tentang Perencanaan Investasi Indonesia.*

Government Regulation 60 of 2012 the Amendment to Government Regulation 10 of 2010 on Forest Determination and Function Change. *Procedure Peraturan Pemerintah Nomor 60 Tahun 2012 tentang Perubahan Peraturan Pemerintah Nomor 10 Tahun 2010 tentang Tata Cara Perubahan Peruntukan dan Fungsi Kawasan Hutan.*

Government Regulation 61 of 2012 on the Amendment to Government Regulation 24 of 2010 concerning the use of forest area. *Peraturan Pemerintah Nomor 61 Tahun 2012 Tentang Perubahan Atas Peraturan Pemerintah Nomor 24 Tahun 2010 Tentang Penggunaan Kawasan Hutan.*

Law1 of 1995 on Limited Liability Company. *Undang Undang Republik Indonesia Nomor 1 Tahun 1995 tentang Perusahaan Terbatas.*

Law 1 of 2004 on State Treasury. *UU No.1 Tahun 2004 tentang Perbendaharaan Negara.*

Law 4 of 2011 on Geospatial Information. *Undang Undang Republik Indonesia Nomor 4 Tahun 2011 tentang Informasi Geospasial.*

Law 5 of 1960 or Basic Agrarian Law re the Basic Provisions concerning the Fundamentals of Agrarian Afforest areairs. *Undang Undang No. 5 Tahun 1960 tentang Peraturan Dasar Pokok-pokok Agraria.*

Law 5 of 1990 on Conservation and Natural Resources and Its Ecosystem. *Undang-Undang Republik Indonesia Nomor 5 Tahun 1990 tentang Konservasi Sumber Daya Alam Hayati dan Ekosistemnya.*

Law 5 of 1994 on Ratification of United Nations Convention of Biodiversity. *Undang-Undang Nomor 5 Tahun 1994 tentang Pengesahan United Nations Convention on Biological Diversity.*

Law 6 of 2014 on Village. *Undang-Undang Nomor 6 Tahun 2014 tentang Desa.*

Law 23 of 1997 on the Management of Environment. *Undang-Undang Nomor 23 Tahun 1997 tentang Pengelolaan Lingkungan Hidup.*

Law 25 of 2004 on National Planning System. *Undang-Undang Nomor 25 Tahun 2004tentang Sistem Perencanaan Pembangunan Nasional.*

Law 25 of 2007 on Investment. *Undang Undang Republik Indonesia Nomor 25 Tahun 2007 tentang Penanaman Modal.*

Law 26 of 2007 on Spatial Planning Law. *Undang Undang Republik Indonesia Nomor Nomor 26 Tahun 2007 tentang Rencana Tata Ruang.*

Law 29 of 2009 on the ammendment of Transmigration Law 15 of 1997. *Undang-Undang Nomor 29 Tahun 2009 Tentang Perubahan Atas Undang-Undang Nomor 15 Tahun 1997 tentang Ketransmigrasian.*

Law 32 of 2004 on Regional Government. *Undang-Undang Nomor 32 Tahun 2004 tentang Pemerintahan Daerah.*

Law 41 of 1999 on Forestry. *Undang-Undang Nomor 41 Tahun 1999 tentang Kehutanan.*

Memorandum of Understanding of 12 Ministries and Institutions on the Acceleration of forest area Stipulation in Indonesia. *Nota Kesepahaman 12 Kementerian/Lembaga tentang Percepatan Pengukuhan Kawasan Hutan.*

Ministry of Agrarian/National Land Agency 2 of 1999on Location Permit. *Peraturan Menteri Agraria/Kepala BPN Nomor 2 tahun 1999 tentang Izin Lokasi.*

Ministry of Agrarian/National Land Agency 3 of 1999 on Delegation of Authority to Grant and Cancellation Decision Granting State Land Rights. *Peraturan Menteri Negara Agraria/Kepala Badan Pertanahan Nasional Nomor 3 Tahun 1999 tentang Pelimpahan Kewenangan Pemberian dan Pembatalan Keputusan Pemberian Hak Atas Tanah Negara.*

Ministry of Agrarian/National Land Agency Regulation 5 of 1999 about Guidelines on Resolution for Communal Land Rights of Indigenous People. *Peraturan Menteri Agraria/Kepala BPN no 5 tahun 1999 tentang Pedoman Penyelesaian Masalah Hak Ulayat Masyarakat Hukum Adat.*

Ministry of Agriculture Decision 786 of 1996 on Plantation Business Licence. *Keputusan Menteri Pertaniannomor.786/Kpts/KB.120/10/96 tentang Perizinan Usaha Perkebunan.*

Ministry of Agriculture Regulation 98 of 2013 on the Guideline of Plantation Business Licence. *Peraturan Menteri Pertanian Nomor 98/Permentan/OT.140/9/2013 tentang Pedoman Perizinan Usaha Perkebunan.*

Ministry of Forestry Decision 340 of 2003 on the Guideline of Implementing National Programme on Forest and Land Rehabilitation. *Keputusan Menteri Kehutanan Nomor 340/Menhut-V/2003 tentang Petunjuk Pelaksanaan Gerakan Nasional Rehabilitasi Hutan dan Lahan.*

Ministry of Forestry Regulation 23 of 2007 on the procedures for applying for a timber utilization permit in community-based plantation forest. *Peraturan Menteri Kehutanan P. 23/Menhut-II/2007 tentang Tata Cara Permohonan Izin Usaha Pemanfaatan Hasil Hutan Kayu Pada Hutan Tanaman Rakyat Dalam Hutan Tanaman.*

Ministry of Forestry Regulation 25 of 2006 on the Guideline of Evaluating National Programme on Forest and Land Rehabilitation. *Peraturan Menteri Kehutanan Nomor P.25/Menhut-II/2006 tentang Pedoman Penilaian Pelaksanaan Gerajan Nasional Rehabilitasi Hutan dan Lahan (GNRHL/GERHAN)*

Ministry of Forestry Regulation 26 of 2012 on the guideline of using title forest. *Peraturan Menteri Kehutanan Nomor P.26/Menhut-II/2005 tentang Pedoman Pemanfaatan Hutan Hak.*

Ministry of Forestry Regulation 30 of 2012 on the forest Products Administration from Title Forest. *Peraturan Menteri Kehutanan P.30/Menhut-II/2012 tentang 2012 tentang Penatausahaan Hasil Hutan Yang Berasal dari Hutan Hak.*

Ministry of Forestry Regulation 35 of 2007 on Non Timber Forest Products. *Peraturan Menteri Kehutanan Nomor: P. 35/Menhut-II/2007 tentangHasil Hutan Bukan Kayu.*

Ministry of Forestry Regulation 37 of 2007 on Community Forestry. *Peraturan Menteri Kehutanan P.37/Menhut-II/2007tentang Hutan Kemasyarakatan.*

Ministry of ForestryRegulation 39 of 2013 on Community Empowerment through Forestry Partnership. *Peraturan Menteri Kehutanan P. 39/Menhut-II/2013 tentang Pemberdayaan Masyarakat Setempat melalui Kemitraan Kehutanan.*

Ministry of Forestry Regulation 44 of 2012 on the Confirmation of Forest Area. *Peraturan Menteri Kehutanan Nomor P.44/Menhut-II/2012 tentang Pengukuhan Kawasan Hutan.*

Ministry of Forestry Regulation 49 of 2008 on village forest. *Peraturan Menteri Kehutanan P.49/Menhut-II/2008 tentang Hutan Desa.*

Ministry of Forestry Regulation P.50 of 2010 on the Method of Granting and Extending The Work Area of Forest Timber Products Utilisation Business Licenses (IUPHHK) In Natural Forests, Ecosystem Restoration, or Industrial Plantation Forests. *Peraturan Menteri Kehutanan Nomor: P.50/Menhut-Ii/2010 Tentangtata Cara Pemberian dan Perluasan Areal Kerja Izin Usaha Pemanfaatan Hasil Hutan Kayu (IUPHHK) dalam Hutan Alam, IUPHHKRestorasi Ekosistem, atau IUPHHK Hutan Tanaman Industri Pada Hutan Produksi.*

Ministry of Forestry Regulation P.53 of 2011 on the second amendment of MoFor Regulation P.49 of 2008 on village forest. *Peraturan Menteri Kehutanan P.53/Menhut-II/2011 tentang perubahan kedua atas peraturan menteri kehutanan nomor p.49/menhut-ii/2008 tentang hutan desa.*

Ministry of Forestry Regulation P.62 of 2013 on the Amendment of MoFor Regulation P.44 of 2012 on the Inaugural of state forest area. *Peraturan Menteri Kehutanan P.62 Menhut-II/2013 tentang Perubahan atas Peraturan Menteri Kehutanan Nomor P.44/Menhut-II/2012 tentang Pengukuhan Kawasan Hutan.*

Ministry of Forestry Regulation P.62 of 2011 on Guidelines for Different Types of Forest Development Business License for Logging on Industrial Forest Concession(IUPHHK-HTI). *Peraturan Menteri Kehutanan Nomor P.64/MENHUT-II/2011 Tahun 2011 tentang Pedoman Pembangunan Hutan Tanaman Berbagai Jenis pada Izin Usaha Pemanforest areaatan Hasil Hutan Kayu pada Hutan Tanaman Industri (IUPHHK-HTI)*

Ministry of Forestry Regulation P.64 of 2011 on the cancelation of Ministry of Forestry Regulation P.62 of 2011. *Peraturan Menteri Kehutanan Nomor P.64/MENHUT-II/2011 Tahun 2011 tentang Pencabutan Peraturan Menteri Kehutanan Nomor P.62/Menhut-Ii/2011.*

Ministry of Forestry Regulation P.56 of 2006 on the guideline for National Park Zonation. *Peraturan Menteri Kehutanan Nomor P.56/Menhut-II/2006 tentang Pedoman Zonasi Taman Nasional.*

Ministry of Forestry Regulation P.44 of 2012 on the Inaugural of Forest Area. *Peraturan Menteri Kehutanan Nomor P.44/Menhut-II/2012 tentang Pengukuhan Kawasan Hutan.*

National Constitution 1945. *Undang-Undang Dasar tahun 1945.*

Presidential Instruction 6 of 2013 on New License Suspension and Improving Governance of Primary Forest and Peatlands. *Instrusksi Presiden Nomor 6 Tahun 2013 tentangPenundaan Pemberian Izin Baru Dan Penyempurnaan Tata Kelola Hutan Alam Primer dan Lahan Gambut.*

Presidential Regulation 10 of 2012 on the amendment of Presidential Regulation 54 of 2009. *tentang Perubahan Atas Peraturan Presiden Nomor 54 Tahun 2009 tentang Unit Kerja Presiden Bidang Pengawasan dan Pengendalian Pembangunan.*

Presidential Regulation 62 of 2013 on Agency of the Management of Greenhouse Gas Emission Reduction of Deforestation, Forest Degradation and Peatland. *Peraturan Presiden Nomor 62 Tahun 2013 tentang Badan Pengelola Penurunan Emisi Gas Rumah Kaca dari Deforestasi, Degradasi Hutan dan Lahan Gambut.*

Presidential Regulation 54 of 2009 on Presidential Task Force Unit for Supervision and Management of Development. *Peraturan Presiden Nomor 54 Tahun 2009 tentang Unit Kerja Presiden Bidang Pengawasan dan Pengendalian Pembangunan.*

The Joint Decrees of the Coordinating Ministry of Welfare, Coordinating Ministry of the Economy and Coordinating Ministry of Politics and Security 09 of 2003, 16of 2003,08 of 2003. *Keputusan Bersama Menteri Koordinator Bidang Kesejahteraan Rakyat, Menteri Koordinator Bidang Perekonomian dan Menteri Koordinator Bidang Politik dan Keamanan No. 09/Kep/Menko Kesra/III/2003, No. Kep.16/M.EKON/03/2003, No. Kep.08/MENKO/POLKAM/III/2003 tentang Pembentukan Tim Koordinasi Perbaikan Lingkungan Melalui Rehabilitasi dan Reboisasi Nasional.*

Land Use Policy 47 (2015) 408–425

Contents lists available at ScienceDirect

Land Use Policy

journal homepage: www.elsevier.com/locate/landusepol

The regime complex for tropical rainforest transformation: Analysing the relevance of multiple global and regional land use regimes in Indonesia

Muhammad Alif K. Sahide[a,b,*], Dodik Ridho Nurrochmat[c], Lukas Giessen[d]

[a] Chair of Forest and Nature Conservation Policy, University of Göttingen, Germany
[b] Laboratory on Forest Policy and Entrepreneurship, Faculty of Forestry, University of Hasanuddin, Indonesia
[c] Forestry Faculty, Bogor Agricultural University, Indonesia
[d] Chair of Forest and Nature Conservation Policy, University of Göttingen, Germany

ARTICLE INFO

Article history:
Received 18 June 2014
Received in revised form 10 April 2015
Accepted 29 April 2015

Keywords:
International agreements
International forest regimes
International land use policies
Native forests
Agroforestry
Rubber plantations
Palm oil plantations
Relevance of international regimes
Problem-dimension
Actor-dimension

ABSTRACT

Increasingly, international agreements, treaties, and conventions address tropical deforestation and specific causes of rainforest transformation. Such international regimes may aim to conserve natural forests or to support the exploitation of lands for a particular purpose, e.g. agricultural commodities. This paper's goal is to map the international regime complex for rainforest transformation by identifying those global and regional regimes that are relevant for tropical rainforest transformation systems in Indonesia. The relevance of international regimes for rainforest transformation is assessed based on a novel methodology using two criteria: First, international regimes may be relevant in Indonesia because the core problem they address exists in Indonesia (problem dimension). Second, specific international regimes may be made relevant by the actions of specific actors and their coalitions in pursuit of their interests (actor dimension). To achieve the study's goals we used (1) content analysis of international policy documents and treaty texts, (2) expert interviews with key informants in Indonesia, and (3) own field observations in current Indonesian land use politics. The results indicate that regimes dealing with biological diversity, climate change, trade in endangered species, wetland management, international tropical trade, illegal logging, Southeast Asian (ASEAN) forest and environment, the Asia Pacific free trade, and the southeast Asian rubber trade regime are relevant in both the problem and actor dimensions. The regimes concerned with desertification and international forestry research are relevant only in the actor dimension. In contrast, the following regimes are relevant only in the problem dimension: indigenous rights, forest certification, palm oil certification, agroforestry certification, and international rubber trade. We discuss our results considering global and regional regimes relevant to Indonesian rainforest transformation systems. We conclude that regimes relevant solely due to the engagement of domestic actors and not in respect of the problem dimension are unlikely to maintain their relevance in the long term. In the short term, however, they have the potential for creating substantial political benefits for the actors using them. In contrast, regimes with high problem relevance but low actor backing are unlikely to even enter the political agenda, and therefore will have only limited impact. This suggests that future research should consider that, whilst relevance in the problem dimension is necessary to establish regime influence, it alone does not suffice. We can expect global and regional regimes to have an influence only when powerful actors back them in their agenda setting, domestic policy formulation and actual implementation.

1. Introduction

Indonesian tropical forest transformation was driven largely by the log export boom in the early 1960s and 1970s (Dauvergne, 1998), the rubber boom in the 19th and 20th centuries (Feintrenie and Levang, 2009) and the recent palm oil boom that began in the 20th century and continues today (McCarthy and Zen, 2010). These had negative impacts on forests and involved greenhouse gas

* Corresponding author at: Busgenweg 3, 37077 Goettingen, Germany. Tel.: +49 17682960836; fax: +49 551393413.
E-mail addresses: msahide@gwdg.de, alif.mksr@gmail.com (M.A.K. Sahide), dnrochmat@yahoo.com (D.R. Nurrochmat), lgiesse@gwdg.de (L. Giessen).

http://dx.doi.org/10.1016/j.landusepol.2015.04.030

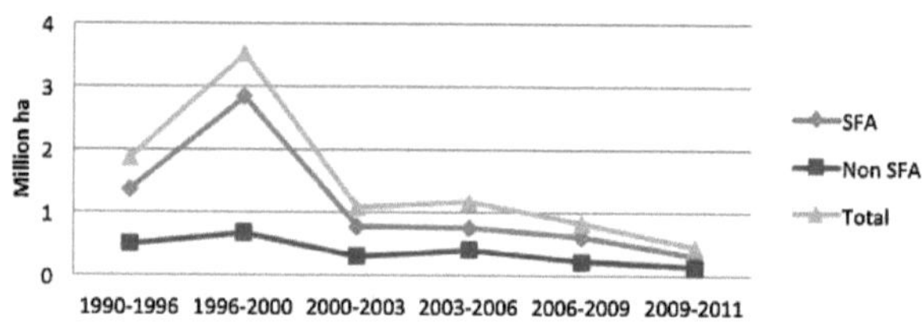

Fig. 1. Deforestation rate in Indonesia from state forest areas (SFA), non-SFAs and total (MoFor, 2012).

emissions and the fragmentation and pollution of habitats (Fitzherbert et al., 2008). In parallel to these recent developments in Indonesia, and over the past two decades, multiple international global and regional forest-related policy initiatives also referred to as "international regimes", have addressed tropical deforestation and its causes. These regimes have had the purpose of influencing domestic policies (Bernstein and Cashore, 2012; Giessen, 2013a,b). In cases where multiple, competing international regimes exist and are loosely coordinated, Keohane and Victor (2011) speak of an international regime complex.

International regimes have been investigated exhaustively for about two decades (Sprinz, 2005) and multiple forest-related regimes have been analysed extensively already (Humphreys, 2006; McDermott, 2012; Smouts, 2008; Rayner et al., 2010; Edwards and Giessen, 2014). However, the domestic relevance of these regimes, as well as their implementation in specific national settings, has been neglected in research, with a few exceptions (e.g. McDermott et al., 2010). This research has been fragmented into various topics, such as regime effectiveness (Young, 2001; Hovi et al., 2003) or regime effects (Lindstad and Solberg, 2010; Lindstad, 2014). A point of criticism by Lindstad and Solberg (2010) is that the concepts of regime effects and effectiveness are very difficult to apply for methodological reasons. Bernstein and Cashore (2012) take up this critique an simplify this by shifting their focus away from the "effectiveness" of regimes and towards their "influence". All these approaches implicitly suggest that international regimes are actually relevant to their member states. This ignores the options of treating such regimes as symbolic policy only. Hence, what these novel approaches underestimate, however, is that in any given domestic context a regime may or may no be relevant in two ways: firstly, the problem addressed by a particular regime may be pertinent specifically to a given country, leading to the relevance of that regime in a **problem dimension** (Krasner, 1982, 1983). Secondly, regimes may be made relevant in a given domestic context due to their active use by domestic policy actors in order to justify their preferences, even if the problem the regime is addressing does not pertain to the country in question. The latter situation is rooted in interest-based bureaucratic politics theory (Krott, 2005; Peters, 2010), according to which national level bureaucracies may utilise international negotiations, regimes and actors to further their interests domestically (Giessen et al., 2014; Giessen, 2013a, 2012; Hofmann, 2002; Hogl et al., 2009). Here, this is considered to be the **actor dimension** of regime relevance.

There are only a few studies that analyse domestic regime relevance. However, Bernstein and Cashore (2012) show that the regimes' forms are determined by the problem structures and actors, which are very diverse in every country. To explain this we combine two separate theories, regime theory itself and bureaucratic politics theory. Sahide and Giessen (2014) have analysed the legal basis and the problem context of Indonesian tropical rainforest transformation. In addition, Wibowo and Giessen (2012) show the international forest issues relevant in an Indonesian context, a topic also seen in similar sources (Burns and Giessen, 2014; Rahman and Giessen, 2014). Based on bureaucratic politics theory, we take an actor dimension as one core measurement for a regime's domestic relevance. Then we follow the logic of regime theory, that regimes should correspond with the issues around which they were created, and we conceptualize it as the problem dimension as the second core measurement for relevance. Specifically, Abbott (2011) uses an actor dimension for assessing the orchestration of the climate change regime complex.

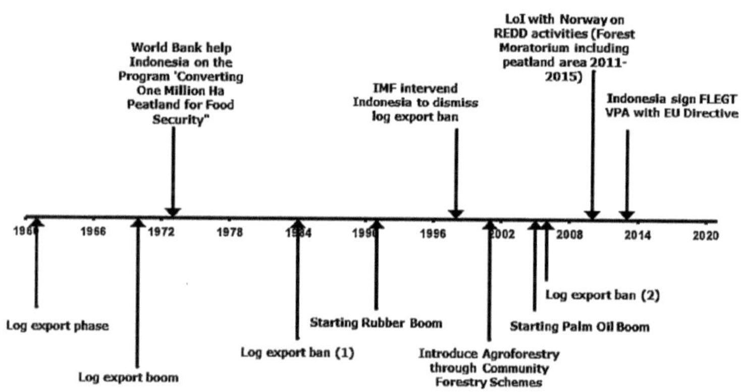

Fig. 2. Timeline of tropical forest transformation in Indonesia (figure based on Dauvergne, 1998; Feintrenie et al., 2010; McCarthy and Zen, 2010; MoA, 2012).

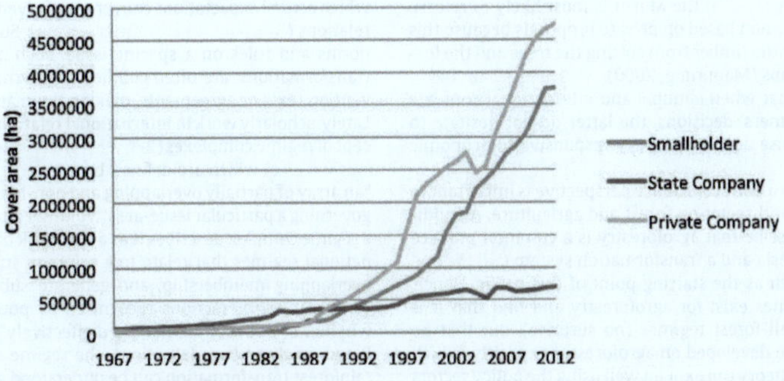

Fig. 3. Indonesian Palm Oil Cover areas (ha) in 1967–2012 by category of producers.
(Source: MoA, 2007; ISPO Commission, 2013a).

Hence, the objective of this article is to (i) identify the international regime complex related to Indonesian tropical rainforest transformation consisting of specific global and regional regimes, and (ii) to demonstrate their relevance for Indonesia by analysing the regimes' problem dimensions and the engagement of important bureaucratic actors in using the regime along an actor dimension. The findings regarding the relevance of global and regional regimes in terms of the problem dimension and the actor dimension will contribute to broader discussions about regime influences and actors' interests in utilising international regimes.

2. Background and theoretical underpining

2.1. Tropical rainforest transformation systems

Tropical rainforest transformation (Steffan-Dewenter et al., 2007; Sahide and Giessen, 2015) has been a persistent issue in Indonesia since the early 1970s (MoFor, 2012).

The data shows that deforestation in Indonesia peaked around 2000 (during the decentralisation era), as seen Fig. 1. An important ecological element of the tropical forest ecosystem in Indonesia is peatland, which makes up about 20.11 million ha. This was part of an ambitious programme by the Soeharto regime in 1973, which involved converting 1 million swamp areas (including peatland) for agriculture or food security (Dauvergne, 1998). This was financed by an international loan through the World Bank and the Asian Development Bank.

Fig. 2 not only confirms that Indonesia has had three main commodities export booms (timber, rubber, and palm oil), but also that it is under the dynamic influence of international regimes. An example would be the IMF's (International Monetary Fund) requirement to lift the Indonesian log export ban of 1998. This case is one example where international regimes govern tropical rainforest transformation systems, which we describe as a change in land use, from native forests to (i) rubber agroforestry, (ii) rubber plantations, and (iii) palm oil plantations.

2.1.1. Transformation to palm oil plantations

The establishment of large-scale oil palm plantations and processing units is an important economic activity that strongly influences the rate of land development and forest clearance in Indonesia (Fitzherbert et al., 2008; Budidarsono et al., 2013; Dhiaulhaq et al., 2014; Miyamoto et al., 2014). Indonesian palm oil, which is the most rapidly expanding tropical crop, supplies about 34% of the world's edible oil consumption (Munadi, 2007), and there are approximately 4.15 million households working on palm oil plantation farms in Indonesia.

Fig. 3 shows a dramatic increase in palm oil plantation area over two decades. There is still a tendency toward an increase at present, with the most recent reading at about 9.23 million ha in 2012. Mostly private companies and smallholders own this, at about 50%

Fig. 4. Indonesian Rubber Cover Area (ha) in 1967–2012 in 1967–2013 by category of producers.
(Source: MoA in Penebar Swadaya, T. P., 2008; MoA in GAPKINDO/Rubber Association of Indonesia, 2014a).

and 40%, respectively (Fig. 3). The MoFor is most likely to release forest areas for plantation based on private proposals because this has a double benefit: the timber from cutting the trees and the forest area for plantations (Manurung, 2000). Feintrenie et al. (2010) support the thesis that when national and international contexts clearly influence farmers' decisions, the latter do not hesitate to transform the land use and appear very responsive to economic opportunities.

Agroforestry from a political science perspective is important to be analysed as a hybrid sector on forest and agriculture. An additional argument must be that agroforestry is a common practice in land use in Indonesia and a transformation system (Sahide and Giessen, 2015), which as the starting point of this paper. Hence, it is ask which regimes exist for agroforestry and find that it is all agriculture and all forest regimes (no surprise), but that no explicit regimes were developed on agroforestry explicitly, which bureaucratic politic theory can explain well using the policy sectors approach (distinct instruments for distinct sectors).

2.1.2. Transformation to rubber plantations

Large-scale rubber plantations started in the 19th century and continued into the 20th (Feintrenie et al., 2010) and strongly influenced the rate of land development and forest clearance in Indonesia (Miyamoto, 2006). Fig. 4 shows how the smallholder benefits the most from this transformation. One can observe that around 2 million households work on rubber plantations (MoA, 2011). This rubber supplies about 26% of global rubber demand (Anwar, 2006) and is categorised both as a forest and a non-forest commodity in Indonesian land policy administration. The competition between rubber and palm oil commodities on the market has motivated people to convert rubber farms to palm oil farms (Republika, 2014a). This is reflected in Fig. 4, which shows that there is no growth in the rubber plantation area after 1997.

2.1.3. Transformation to rubber agroforestry systems

The smallholders' rubber production segmentation is associated mostly with agroforestry that follows very diverse patterns (Joshi et al., 2001; Neyrma et al., 2012). Rubber agroforestry tends to achieve a maximum ecological growth that is similar to that of a virgin topical forest managed by indigenous groups in Indonesia (Feintrenie and Levang, 2009) and developed by means of endogenous innovations (Budiman and Penot, 1997). Therefore, Indonesian rubber agroforestry as reported by Lawrence (1996) is the trade-off between rubber production and the maintenance of diversity. In this way, rubber agroforestry in Indonesia can potentially produce two types of timber: rubber wood and wood from other species found in agroforestry, such as fruit trees (especially durian), soft wood species, and Dipterocarpaceae (Gouyon, 2003).

2.2. Regime complexes, global and regional regimes

Fundamentally, an international regime can be defined as a set of implicit or explicit principles, norms, rules and procedures around which actorsí expectations converge in a given area of international relations (Krasner, 1982; Giessen, 2013a). Such sets of principles, norms and rules on a specific issue, such as tropical rainforest transformations, are often codified in international treaties, convention texts or agreements, making them accessible for analysis. Lately, scholarly work in international relations developed the concept of regime complexes (Raustiala and Victor, 2004; Keohane and Victor, 2011) which are defined by Raustiala and Victor (2004) as "an array of partially overlapping and non-hierarchical institutions governing a particular issue-area". Another alternative definition of a regime complex describes it as a "network of three or more international regimes that relate to a common subject matter, exhibit overlapping membership, and generate substantive, normative, or operative interactions recognised as potentially problematic whether or not they are managed effectively" (Orsini et al., 2013). Based on these two definitions, the regime complex for tropical rainforest transformation can be understood as a network of institutional elements not organised hierarchically, including entire international regimes involved in a wide array of activities associated with the multiple issues regarding rainforest transformation.

In the field of international forest-related cooperation the concept has been identified as being useful for empirical analyses (Giessen 2013a; Rayner et al., 2010). However, forests are subject to influences from multiple policy fields and inter-sectoral interactions (Hogl et al., 2009; Giessen and Krott, 2009; Humphreys, 2009; Krott and Hasanagas, 2006; Rayner et al., 2001). Hence, the concept of a regime complex affecting tropical rainforest transformation systems is proposed in this paper, and accounts for adjacent policy fields that also regulate rainforest transformations in part.

A regime complex dedicated to tropical rainforest transformation might include a number of institutional elements devoted to different issues. These elements might be of global or regional nature. In this context, regional regimes are understood as international regimes between two or more states, with criteria limiting membership on the basis of spatial or other crucial functional proximity, and a lack of global aspirations (Giessen et al., unpublished; for examples see Overdevest and Zeitlin, 2014; Nurrochmat et al., 2014; Derkyi et al., 2013; Jürging and Giessen, 2013; Edwards and Kleinschmit, 2013).

3. Methodology

3.1. Measuring regime relevance: being relevant vs being made relevant

The regime complex is characterized by the unstructured and overlapping each other (Raustiala and Victor, 2004), however, here it is developed a methodology that could be a new offer to the theory that can potentially break the unstructured form by developing common indicators across regimes and regimes' category using bureaucratic politic theory (Krott, 2005; Peters, 2010) and regime (Krasner, 1982, 1983) it self for both dimensions.

Table 1
Indicators to assess the problem dimension of the relevance of international regimes.

Code	Indicator	Scoring method	Points
P1	Indonesian tropical rainforest transformation issues formally fit in with the regime's scope or aims	High match	2
		Medium match	1
		No match	0
P2	The problem intensity and or issues quality of Indonesian tropical rainforest transformation system	High problem	2
		Medium problem	1
		Less problem	0
P3	The regime's scope is compatible with Indonesian priority on their forestry transformation's plan	High priority	2
		Medium priority	1
		Little/Not priority	0

Table 2
Indicators to assess the actor dimension of the relevance of international regimes.

Code	Indicator	Scoring Method	Points
Domestic actors			
A1	Bureaucratic structures reflecting responsibility for a regime	There is a specific autonomous bureaucracy dealing with the regime	2
		There is a specific unit appointed for dealing with the regime, e.g. national focal point, national authority, etc.	1
		There is no special unit appointed for dealing with the regime	0
A2	Bureaucracies' budget allocations for a regime	Substantial budget or contribution for additional activities	2
		Marginal budget allocated according to minimum requirement of a regime	1
		No budget allocated	0
A3	Bureaucracies' legal mandate to implement a regimes	High support by laws (*Undang-Undang*)	2
		Medium support on a legal basis (like government regulation, presidential regulation, presidential instruction) or high support by laws but different interpretation and difficult to implement	1
		No formal mandate or only tacit agreement within the government	0
A4	Actors promoting regime and broadening coalitions is support for a regime	Successful coalition building in support of a regime including at least one state bureaucracy	2
		Successful coalition building in support of a regime among non-sate actor only	1
		No successful coalition building in support of a regime (minimum one promoter plus one coalition partner)	0
International (and regional) actors			
A5	International bureaucracies' in charge of implementing a regime in Indonesia	There is a specific unit of an international bureaucracy on regime implementation only in Indonesia	2
		There is a specific unit of an international bureaucracy on regime implementation in the region, including Indonesia	1
		There is no specific unit of an international bureaucracy on regime implementation in Indonesia	0
A6	International actors promoting regime and broadening coalitions is support for a regime	Successful coalition building in support of a regime including at least one state bureaucracy	2
		Successful coalition building in support of a regime among domestic and international actors	1
		No successful coalition building in support of a regime (minimum one promoter plus one coalition partner)	0

Problem dimension: The problem dimension is used to assess whether international regimes may be relevant in the context of Indonesian tropical forest transformation, because the core problem and issues of the regimes address exist in Indonesia. According to Bernstein and Cashore (2012), international regimes placed in a national context are different in each country and depend on problem structures, actors and the institutional setting. Table 1 provides criteria and indicators for assessing and quantifying the problem dimension.

Actor dimension: Actor dimension is introduced because specific actors may also make international regimes relevant if it is useful for them in the pursuit of their interests. These actors may be public or private actors and internationally or domestically based (Arts, 2012; Arts et al., 2013). However, Krott (2005) argued that it is bureaucracies' interests that determine their actions and that therefore these interests are one of the most important factors in political processes. Hence, regimes may be made relevant by specific bureaucracies in a given country even if the core issue they address might not even exist in this very country. This dimension is rooted in framing theory, according to which issues may be created by actors even if they do not have a strong biophysical grounding (Sadath et al., 2013). Below we provide Table 2 to assess the actor dimension.

Based on the above indicators, which distinguish between "problem dimensions" and "actors' behaviour", we create a matrix that will help us to differentiate the different types of relevance of the regimes in the context of Indonesian tropical rainforest transformation. We use four quadrants to examine the degree of relevance of regimes implemented in Indonesia (see Fig. 5). Quadrant I is the most relevant regime, quadrants II is relevant only in the actor dimension, while quadrant III represents relevance only in the problem dimension, and quadrant IV according to our measure is not relevant in Indonesia. We measure the quality of both dimensions by using quantification, where the range 0–3 points is not relevant and >3–6 points is relevant for problem dimension, and the range 0–6 points is not relevant, and >6–12 points is relevant for actor dimension.

3.2. Materials and method

Regimes and its domains of Indonesian tropical rainforest transformation, which come from ecological land use research and their recent finding adding analytical value by providing more land use categories than 'forest' and 'non forest', are selected through dual approach such as:

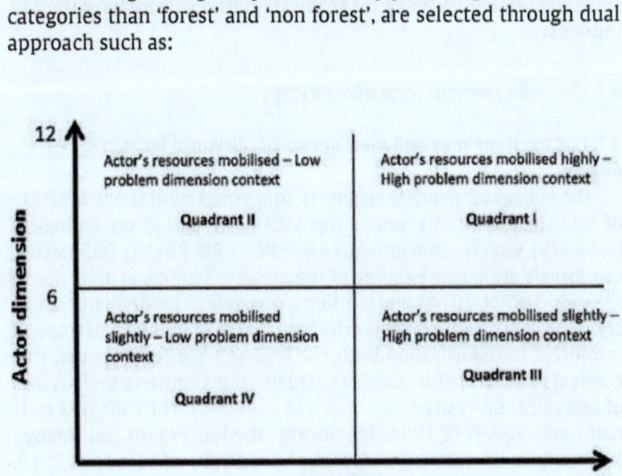

Fig. 5. Analytical matrix for assessing the relevance of global and regional regimes for rainforest transformation along actor and problem dimensions in Indonesia.

1. Deductively, the authors oversee all forest related regimes in the international forest regime complex (Giessen, 2013a,b) and conducted literature review on palm oil, rubber, and agroforestry. By this a pre-selection of potentially relevant (either actor or problem dimension) regime was identified. The main documents come from international organisations and national ministries related to tropical forest transformation. Specific literature from (Wibowo and Giessen, 2012) that has identified international forest issue related in Indonesia is also used. This is inline with (Sadath and Krott, 2012) that issues can be a starting point in policy analysis.
2. Inductively, it is delivered interviews and observations with people with a broad knowledge of Indonesian forest transformation systems that starting from foreign cooperation division of related ministries in in charge of tropical forest transformation in Indonesia such as the Ministries of Forestry, Agriculture, Environment, Industry and Trade.

Observations and interviews were conducted in Indonesia between December 2013 and February 2014, and to increase reliability we carried out a second field investigation between October 2014 and November 2014. We then measured the relevance of the individual regimes based on the problem and the actor dimensions. Below are the three main methods used to support this research:

1. Literature review is basically a qualitative content analysis that is a flexible method for analysing text data (Hsieh and Shannon, 2005; Neuman, 2005).
2. The chosen interviews were delivered to develop a causal explanation for our theory (based on Mosley, 2013).
3. Observations were conducted based on Van Evera (1997), meaning that first we took the theory to the field, made observations to generate data without imposing an external stimulus on the situation, and then we questioned whether the observations were compatible with our theory.

4. Results

Based on the indicators, it is analysed the relevance by using the code, which is attributed to the indicators to show the quantity of points and describe the character of each dimension (see Tables 1 and 2). Table 3 provides the results the relevance analysis of the regime complex for tropical rainforest transformation in Indonesia.

4.1. Regimes relevant for native forests

4.1.1. Global regimes and their degree of relevance for native forests

The biological diversity regime is concerned with the protection of biological diversity under the CBD (Convention on Biological Diversity), which came into force in 1992 (CBD, 2015a). Indonesia's rainforests are home to some of the greatest biological diversity in the world (RAN, 2014), and it is known as a mega biodiversity country (MoE, 2012), which shows the high match of Indonesian tropical rainforest transformation with the regime's scope. However, this tropical forest transformation into plantations causes a serious loss of biological diversity (Fitzherbert et al., 2008). The CBD was ratified under Law 5 of 1994. Specifically, the Rotterdam and Nagoya Conventions are ratified under Law 10 of 2013 and Law 11 of 2013. The ministries responsible for forestry and environment were given the responsibility of implementing these laws as the main actors as well as the National Competent Authority for CBD secretariat in Montreal (MoE, 2013). For example, the special state budget for the environment was increased by 20% in 2012. This represents IDR 479,730,000,000 allocated to 442 districts under the MoE budget (MoE, 2011). The MoFor also allocated IDR 1355.33 billion in 2014 for biodiversity conservation (this is mentioned in MoFor Regulation P.51 of 2010). In the national forestry plan (in place until 2030), the government directed 26,819,383 ha for a conservation area to include biological diversity protection (MoFor Regulation 45 of 2011). This conservation area has the most restrictions against transformation into other land uses (Sahide and Giessen, 2015). There is no direct CBD secretariat in Indonesia, however, CBD has developed coalition with other domestic and international actors through Memoranda of Cooperation and Joint Work Programmes (CBD, 2015b), and it is observed that a lot of CBD's activities that directly with MoE and MoFor. MoFor has developed cooperation with other public and private actors on preserving biodiversity under MoFor Degree 390 of 2003.

The wetlands regime has assumed wetland preservation as its core issue under RAMSAR (The Convention on Wetlands of International Importance) since 1975 (RAMSAR, 2014a). There are 7 conservation areas that have been designated as RAMSAR sites in Indonesia (RAMSAR, 2014b). Indonesia has 20.6 million ha of peatland, representing 10.8% of the total land surface (WI-Indonesia 2014). There is the threat of destruction of mangroves and unsustainable oil palm expansion in Indonesia's wetlands, especially in peatland areas (WI, 2014b). Several important actions have been attempted after the RAMSAR conventions were ratified under the Indonesian Presidential Decision 48 of 1991, such as forming working groups on RAMSAR (based on MoFor decree 22/IV-KK/2007) It is observed that MoFor engages other international actors to support peatland protection, for example, Wetland International (WI) and its branch, Wetland International Indonesia Programme (WI, 2014a; WI, 2014b). The other example, a joint initiative of the Australian and Indonesian governments under the Kalimantan Forests and Climate Partnership provide a huge budget, which is around A$37 million for four years to addressed good peatland management in Kalimantan (Evans, 2014). RAMSAR is also observed developed initiative for such regional cooperation which involves some Indonesian bureaucracies and International actors (RAMSAR, 2014c). For the RAMSAR budget submission, the government have allocated around US$ 4500 each year (Murniningtyas et al., 2007). There is also the implementation of a moratorium programme for all activities on peatland areas by 2015 (President Instruction 6 of 2013), and the government has allocated 28.4 ha of peatland to be protected until 2030 (MoFor Regulation 45 of 2011). Domestic private actors like Riau Pulp and Paper Group has been evaluated and resulted a good peatland management performance (Rainforest Alliance, 2015) and was appreciated by the MoFor.

The endangered species trade regime was formed in 1963 and aimed to ensure that international trade in wild animal and plant species did not threaten their survival under CITES (Convention on International Trade in Endangered Species of Wild Fauna and Flora), of which Indonesia became a member in 1978 (CITES, 2014; MoFA, 2014a). Indonesia has ratified CITES under Presidential Decision 43 of 1978. The ratification is also supported by the Law 5 of 1990 on Biodiversity Conservation and its Ecosystem. The country faces illegal trade in endangered species and has registered 1053 flora and 1384 fauna species on Appendices I and II of the CITES Convention (MoFor Regulation P.51 of 2010). The Ministry of Forestry (which was integrated with Ministry of Environment in October 2014) has placed the protection of endangered species among other priorities of the Forestry Plan 2011–2030, like the commitment to provide conservation funds for 236 species (MoFor Regulation 49 of 2011). The Ministry of Marine and Fishery also provide funds for this species protection. It is observed that those bureaucracies have developed partnership with private actors, particularly on CITES campaign. Moreover, CITES appointed Directorate General of Forest Protection and Nature Conservation of MoFor as the repre-

Table 3
The relevance analysis of the regime complex for tropical rainforest transformation in Indonesia.

No	Regime complex	Points collected									Qd.
		P1	P2	P3	A1	A2	A3	A4	A5	A6	
1.	The biological diversity regime	2	2	2	1	1	2	2	0	2	I
2.	The wetlands regime	2	2	2	1	1	1	2	0	2	I
3.	The endangered species trade regime	2	2	2	1	1	2	2	0	2	I
4.	The climate change regime	2	2	2	2	2	2	2	2	2	I
5.	The illegal logging regime	2	2	2	1	2	1	2	2	2	I
6.	The International tropical timber trade regime	2	2	2	1	1	1	2	0	2	I
7.	Regimes that combat desertification	1	1	0	1	2	1	2	0	2	II
8.	The forest certification regime	2	2	0	0	0	0	2	0	2	III
9.	The Indigenous rights regime	2	2	2	0	1	1	1	0	2	III
10.	The International forestry research regime	1	1	1	1	2	1	2	2	2	II
11.	The Southeast Asian free trade regime	2	2	2	1	2	2	2	2	2	I
12.	The Asia Pacific free trade regime	2	2	2	1	2	2	2	0	2	I
13.	The Southeast Asian forest and environmental regime	2	2	2	1	1	2	1	1	2	I
14.	Asian and Pacific forestry share an information regime	2	1	0	1	0	0	0	0	2	IV
15.	Tropical rainforest countries' information sharing regime	2	1	0	1	0	0	1	0	0	IV
16.	Agroforestry certification regime	2	2	1	0	0	0	2	0	2	III
17.	International rubber trade regime	2	2	2	0	0	0	2	0	2	III
18.	The Southeast-Asian rubber trade regime	2	2	2	1	2	1	2	1	2	I
19.	Palm oil certification regime	2	2	2	0	0	0	2	2	2	III

Remark:
Qd.: Quadrant.
Code for Problem Dimension: P1, P2, P3; 0–3 points: not relevant, >3–6 points: relevant.
Code for Actor Dimension: A1, A2, A3, A4, A5, A6; 0–6 points: not relevant, >6–12 points: relevant.

sentative contact for CITES. The MoFor and the Indonesian Research Institution (LIPI) have been appointed as the managing and the scientific authorities in charge, respectively (Government Regulation 8 of 1999). Besides providing a national regular budget as part of its national plan, Indonesia also contributes between US$ 9000 and US$ 18,000 annually to CITES (Murniningtyas et al., 2007). On the cooperation with other actors, CITES has been accepted as permanent observer status in regular sessions of the WTO Committee on Trade and Environment (CTE) and a joint programme with ITTO on Sustainable Trade and Management of Timber Species (Prihadi, 2013).

The climate change regime: The Indonesian archipelago is vulnerable to the impacts of climate change, resulting in changes in the weather and natural disasters (Allison et al., 2009; Measey, 2010). Therefore, the climate change regime examines efforts on climate change mitigation and adaptation under the UNFCC (United Nations Framework Convention on Climate Change) conventions, which in these problem dimensions are divided into REDD+ (reducing emissions from deforestation and forest degradation and the role of conservation, sustainable management of forests and enhancement of forest carbon stocks in developing countries) and CDM (clean development mechanism) agreement under the Kyoto Protocol of 1997 (UNFCC, 2014). Since the President Yudhoyono announce their target on reducing emission both in G20 meeting and COP-15 and was followed by Indonesia's submission to UNFCCC on 19 January 2010, which voluntary reduce GHG emission for 26% from Business As Usual level by year 2020 and up to 41% with international support (Purnomo et al., 2013), we can easily observed Indonesian national plan that put climate change on their national priority (like MoFor 49 of 2011, and the document of Ministry of National Planning (2014) on National Action Plan of Climate Change Adaptation), and multiple domestic and international actors was involved on this regime implementation in Indonesia. Moreover, special international structure was formed only to coordinate Indonesian REDD+ activities in United Nations systems, which is called UNORCID or the United Nations Office for REDD+ Coordination in Indonesia (UNORCID, 2014). The CDM is difficult to implement due to the limiting factors of the CDM area requirements, which indicate that the land in question should be land that was degraded before 1990 (Wibowo, 2008). The same problem is present in the case of REDD+, for which there is still no final concept for a multilateral agreement. Nevertheless, Indonesia has shown some development on REDD+ issues since the COP 13 in Bali in 2007 (Interview 4, 2013). The core domestic bureaucracy is the NCCC (National Council on Climate Change), which is led directly by the President. It is in charge as the focal point of the UNFCC and leads the IFCA (Indonesian Forest Climate Alliance), which consists of multiple bureaucracies and civil societies in Indonesia and is supported by multiple international donors (MoFor 2014d). Another special bureaucracy is REDD+ Agency (*Badan Pengelola* REDD), which existed under Presidential Regulation 62 of 2013. At least two laws, two government regulations, three presidential regulations, three presidential decrees and four MoFor regulations (from 1994 to 2013) have been identified as supporting REDD+ and CDM activities in Indonesia (Siregar and Ridwan, 2013). The MoFor allocated a budget for environmental management and climate change research of around IDR 12.30 billion in 2014 and for the standardisation for the environment and climate change of about IDR 31.54 billion in 2014 (MoFor Regulation P.51 of 2010).

The illegal logging regime is a regime initiated by domestic actors that engages countries multilaterally to combat illegal logging and to provide legal timber from sustainable timber management. Indonesia faced around 3260 cases of illegal logging in 2005–2009 and it is allocated more than IDR 66.10 billion for forest protection (including the fight against illegal logging) in 2014 (MoFor Regulation 51 of 2010). Timber legality is a state instrument to approve formally the support of the supply and procurement of legal and sustainable forest products and the setting of conditions that allow the harvest and trade of forest products that comply with applicable local, national, and international legal requirements (Forest Legality Alliance, 2014). This achievement is a result of Voluntary Partnership Agreement (VPA) on the Forest Law Enforcement, Governance and Trade (FLEGT) between Indonesia and the European Union (FLEGT-VPA) which has been sign in 2013 and ratified in 2014 (EU FLEGT Facility, 2015). Following the agreement, Indonesia started to develop a mandatory timber legality scheme called the Timber Legality Verification System (SVLK) (MoFor 2014). There is no special unit from EU for this particular FLEGT implementation in Indonesia, however, special program which is called MFP (Multi-Stakeholder Forestry Programme) is delivered on Indonesian SVLK issues as the international program from collaboration with United Kingdom aid and MoFor ha. A lot of domestic actors through multi-stakeholders consultations have been involved on MFP activities (Interview 7, 2014). The MoFor and MoT have developed regulations to support SVLK, like MoFor Regulation 42 of 2013 and MoT Regulation 64 of 2012. In the beginning, a budget for forest management and timber legality verification was guaranteed by the government (MoFor Regulation 38 of 2009). Furthermore, SVLK was put into place as the forestry strategic national plan (MoFor Regulation 45 of 2011). The MoFor and MoI spent a large budget to develop the Timber Legality Verification System (SVLK). For example, 5 billion rupiahs was provided by the MoFor to support the free SVLK certificate for the smallholder wood industry (Detik, 2014), and the MoI provided 100 free SVLK certificates for the smallholder wood industry (Republika, 2014b).

The International tropical timber trade regime: The ITTA (International Tropical Timber Agreement) regime was adopted in 1983 and went into effect in 1985. It is operationalized by the ITTO (International Tropical Timber Organization). This regime evolved from the ITTA 1983 to the ITTA 1994 and on to the ITTA 2006, which aims to promote the expansion and diversification of international trade in tropical timber from sustainably managed and legally harvested forests. This agreement is considered to be a legally binding document by its signatories (UNCTD, 2006). Indonesia's exports of timber and timber products settled at around US$10 billion last year (Yulisman, 2014). Indonesia still contributes budget between US$ 200,000 and US$ 400,000 yearly to the ITTA for its membership (Murniningtyas et al., 2007 and Interview 2, 2014). Indonesia has taken more ITTO's projects than those budget submitted to the ITTO projects (Interview 2, 2014). It is observed that ITTO addressed project by working not only with state actors but also non-government organizations. Those activities are coordinated by the Centre for Foreign Cooperation of MoFor as the national focal point for ITTO (Interview 2, 2014; NRDCF, 2014). Indirectly, regulation has been set by the governments in the form of MoFor and MoT regulations in order to implement this regime. This regime has seen its influence be reduced due to Indonesia's start of the log export ban for reasons having to do with the restructuring of the forest industry (Interview 1, 2014) and there is no (direct) special plan to apply this regime in the national forestry plan, until 2030, in MoFor Regulation 45 of 2011, however, Indonesia have large and long term timber export target.

Regimes that combat desertification: These regimes were formed in 1996 and were developed under the UNCCD (United Nations Convention to Combat Desertification). They aimed to develop research, techniques, and new approaches, including international support, to overcome land degradation (UNCCD, 2014). Although land degradation is one of the issues in this regime but desertification as the core issues and Indonesia did not suffer desertification. However, Maplecroft reports that Indonesia is the third country worldwide in terms of the extent of deforestation and degradation, after Nigeria and North Korea (NG Indonesia, 2011). The MoFor targeted 2.5 million ha of degraded land to be rehabilitated by 2014 (MoFor Regulation 51 of 2010). In reality, this regime is for countries experiencing serious drought and/or desertification, particularly in Africa, but the UNCD was also approved by the Indonesian Presidential Decree 135 of 1998. There is no (direct) plan for desertification issues in the national forestry plan based on the MoFor Regulation 45 of 2011. The Directorate General for Land Rehabilitation and Social Forestry of the MoFor is the national focal point, or the representative, of the UNCCD (Murniningtyas et al., 2007). A huge budget was allocated by the MoFor for land rehabilitation on 2,009,881 ha annually in 2003–2008 and about IDR 61.85 billion in 2014 for forest rehabilitation programmes (MoFor Regulation 51 of 2010). It is observed that different actors from private and non-governmental organizations involve on the national land rehabilitation program based on the MoFor Degree 3 of 2004 guideline. Based on the Ministry of Forestry report for UNCCD, there are many programs that have been delivered in Indonesia under UNCCD convention (Prakoso, 2002). It is observed that international actors also collaborate with state and non state actors on promoting the fighting for deforestation in Indonesia

The forest certification regime: Indonesia has a production forest area of about 75,869,485 ha (MoFor 2013) and a private forest area of around 1,568,415.63 ha (MoFor 2005). This produced around 49,258,255 m^3 in logs in 2012 (MoFor 2013), which could be potentially certified. Certification relies on evaluations and audits using an evolving set of standards, and as these systems mainstream, many analysts expect these standards to erode and the certification organisations to lose legitimacy as they are captured by the most powerful stakeholders (Klooster, 2010). The FSC (Forest Stewardship Council), formed in 1994, LEI (*Lembaga Ekolabel Indonesia*), founded in 1998 and the PEFC (Programme for the Endorsement of Forest Certification), which was founded in 1999, are the most involved in Indonesian forest management. This is intended to reassure consumers that the products they are purchasing have been produced to the highest ethical and environmental standards, which are those that have been recognised by MoFor for voluntary certification (FSC, 2014; LEI, 2014; PEFC, 2014a). PEFC is a rather new operation in Indonesia. The Indonesian Forestry Certification Cooperation (IFCC) has been approved as a National Governing Body member of the PEFC Council for Indonesia. The IFCC involves MoFor staff on its multi-stakeholders' activities (Interview 5, 2014 and IFCC, 2014). The regime scope from all these voluntary certification schemes is matched to the Indonesian tropical transformation system. Due to the nature of voluntary scheme, there is no state unit as representative regime, no formal bureaucracies involved, no mandate, as well as no budget allocated. A MoFor representative, the National Standardisation Agency (NSA), and representatives from universities and NGOs were actively involved in the initiation of an LEI development in 1999 that was called "the LEI Certification Working Group" (Interview 6, 2014). However, the involvement of MoFor representatives and the bureaucracy staff with these voluntary certification organisations is only individual or informal. Furthermore, there is no formal target in the national forestry plan for accessing this voluntary certification based on MoFor Regulation 45 of 2011.

The indigenous rights regime advocates indigenous rights via FPIC (Free prior and informed consent) principles of forest management (Humphreys, 2006). The United Nations Declaration on the Rights of Indigenous People (UNDRIP) contains the most complete description of these and of the obligations of states regarding the rights of indigenous peoples, including their rights to own and use their traditional territories. The Ministry of Social Affairs (MoSA) formally recognised around 229,479 households of remote Indonesian indigenous communities in 2658 locations (MoS, 2014). At the same time, the Indigenous Peoples' Alliance of the Archipelago (AMAN) as the main actor for indigenous issues in Indonesia estimated that between 50 and 70 million people could be categorised as having indigenous rights in Indonesia (AMAN, 2011). In addition, 80% of the customary forests have been managed by forest concession (Antara, 2014). Actually, Sahide and Giessen (2015) report that on Indonesian national forestry plan, the government Indonesia has open the possibility for indigenous people to manage their forest through different formal scheme such as *hutan desa* (village forest), *hutan kemasyarakatan* (community forest) and *hutan adat* (customary forest). The Community Forestry Communication Forum (FKKM) or the alliance of civil society (including representative from MoFor) for community forest argue that these schemes should respect and recognize the indigenous rights (Observation 3, 2013).

The MoFor has a lot of activities working with non-governmental organizations to promote and implement this scheme, appointed unit to serve the scheme, and allocated regular budget only to formally assist the administration procedure (Interview 3, 2013). However, special unit appointed unit only to serve this schemes, not the all substance of the regime's implementation. MoFor still questions the requirements for indigenous criteria through laws and regulations, for example, the implementation of Constitutional Court Decision 35 of 2012 (Interview 3, 2013; Sahide and Giessen, 2015) and there are no laws and rules to implement FPIC principles directly, (A3). Indonesia is a signatory to the UNDRIP, which was adopted by Indonesian Government, but was not implemented due to the competing interests among Indonesian actors and their definitions of what is indigenous (Observation 3, 2013). Therefore, there is no specific unit or secretariat appointed from UNDRIP to implement the regime in Indonesia. However, it is observed that a lot of international actors working with domestic non-state actors to promote FPIC rights.

The International forestry research regime: Indonesia, as an important tropical country, need good quality on forestry researches both on scientific and practical research, and has many forestry research centres, both in the Ministries and Universities, and has allowed international research actors to do research in Indonesia at places such as the Centre for International Forestry Research (CIFOR) and the World Agroforestry Centre (ICRAF), under the Consultative Group on International Agricultural Research (CGIAR). The CGIAR was established in 1971 and today consists of an informal association of 41 public and private sector donors, supporting some 18 international agricultural research centres including in Indonesia. Indonesia has formally facilitated CIFOR/CGIAR by agreeing that Indonesia should become a headquarters for the CIFOR, through Indonesian Presidential Decree 71 of 1993 (MoFA, 2014a). CIFOR and ICRAF coordinate actors throughout their activities in Indonesia and use "action research" or "participatory research" methodology to engage bureaucratic involvement (Observation 2, 2014), however, Clark et al. (2011) found that under the concept of boundary work, CGIAR's research programs that engaging different actors (like farmers and policy makers) on producing useful knowledge for sustainable development is not harming science and even a part of method, but it need multiple forms of knowledge to avoid the extreme politicization of formal knowledge. It is observed that many local and national non-governmental organizations as well as civil society coalitions are involved with CIFOR and ICRAF through action and joint research. MoFor has established a special institution to conduct forestry research, to allocate regularly special budgets and to include this in their national forestry plan as well. However, this domestic institution is independent from international research regime. Indonesia contributes approximately US$ 58,000 annually to CIFOR/CGIAR (Murniningtyas, 2007), and the MoFor provides 300,000 ha for CIFOR's education forest (Iskandar, 1999).

4.1.2. Regional regimes relevant for native forests

The Southeast Asian free trade regime is under ASEAN charter, which formed on 8 August 1967 and has 10 member countries in the Southeast Asian regions, with headquarters in Jakarta (MoFA, 2014a). The agricultural production potential of Indonesia has to be seen in the light of a larger regional cooperation, especially since the cash economy spreads rather quickly in present-day Indonesia (Dequin, 1975). Now ASEAN has agreed to liberalise their market through AFTA (The ASEAN Free Trade Area) and the ASEAN Economic Community (AEC) 2015, which some forest and agroforestry products are included in the agenda (MoFor 2004). Indonesia has ratified the ASEAN Charter through Law 38 of 2008. All sector ministries must support AEC preparation, especially those related to the 12 sectors that are led by the MoFA (Ministry of Foreign Affairs) and MoT. The MoFa has the special Directorate General for ASEAN affairs as a first level bureaucracy, and it has a special budget, for example, for a procurement plan in 2014 of around IDR 953,099,000 (MoFa, 2013). It is observed that MoFA, MoT, MoI, and MoFor have involved many domestic actors to socialize this regime and a lot of international actors assist ASEAN and Indonesia to implement this regime (Observation 5, 2013). The intergovernmental based lead the ASEAN Secretariat has authority to only ensure the compliance of AFTA measures, or the secretariat has no legal authority to enforce compliance. For this purpose, the regime has formed special bureaucracy, which is named AFTA council. It is observed that non-state actors has involved on promoting this regime involving both Indonesia and ASEAN Secretariat, especially form domestic and regional private sectors.

The Asia Pacific free trade regime, under APEC (Asia-Pacific Economic Cooperation), is a forum formed in 1989 for 21 Pacific Rim member economies that seeks to promote free trade and economic cooperation throughout the Asia-Pacific region and that plays an important role in pursuing green growth in the region (APEC, 2014). Indonesia has successfully increased its exports to APEC countries tenfold in the last 22 years (1989–2011). This represents about 78% of total Indonesian exports around the world (Antara, 2014). The recent agreement has the goal of reducing tariffs to a maximum of 5% by 2015 on a negotiated list of 54 environmental goods (APEC, 2014 and ICTSD, 2014). Global free trade jurisdiction under General Agreement on Tariff and Trade – The World Trade Organization is applied to support this regime and Indonesia has approved this agreement through its Law 7 of 1994. The MoT and the MoFA are the core bureaucracies involved with the APEC regimes and will coordinate with other sector-based ministries, such as the MoFor and MoA. The MoFa has a special Directorate General for Asia Pacific and Africa affairs as one of the core bureaucracies apply this regime (MoFa, 2014b). Both international and domestic private actor actively involve on promoting and using the regime.

The Southeast Asian forest and environmental regime: The ASEAN charter was approved by Indonesia through Law 38 of 2008. ASEAN is also concerned with forest and environmental issues among its members. An example of this is the ASEAN Agreement on Transboundary Haze Pollution (ATHP), which is the first regional arrangement in the world that binds a group of contiguous states to tackle transboundary haze pollution resulting from land and forest fires (ASEAN, 2014). Indonesia ratified this treaty due to the fact that most of the recent haze pollution comes from Indonesian islands. The ASEAN Agreement on transboundary haze pollution (ATHP) is strongly related to the tropical forest transformation into palm oil, due to the Indonesian haze pollution caused by the slash-and-burn method of palm oil plantation. Indonesia is the last country to ratify ATHP through Law 26 of 2014. The roles of other ASEAN member states, like Malaysia and Singapore, which suffered from haze pollution, also contributed to a state of affairs, which led Indonesia to finally ratify ATHP. Previously, the MoFor and the MoE had taken mutually opposing positions. The MoE pushed Parliament to ratify ATHP, but the MoFor still argued that haze pollution is a complex issue that could not be resolved by merely implementing ATHP. This opposition was reduced, however, after integrating the MoE and MoFor into the Ministry of Environment and Forestry (MoEFor), (Observation 6, 2013). Furthermore, the ministry responsible for forestry, agriculture, and environment is the core bureaucracy having to do with this regime. Some working groups related to forestry and the environments have been formed under the Meeting of ASEAN Ministers on Agriculture and Forestry (AMAF), as well as ASEAN Ministerial Meeting on the Environment (AMME). For example: (i) The ASEAN Centre for Biodiversity (ACB), formerly called the ASEAN Regional Centre for Biodiversity Conservation (ARCBC), began in 1999 (ACB, 2014); (ii) the ASFN (ASEAN Social Forestry Network), launched by the ASEAN Senior Officials on

Forestry (ASOF) in August 2005, is the first and largest government-driven social forestry network in Southeast Asia (ASFN, 2014); (iii) the ASEAN Regional Knowledge Network on Forests and Climate Change (ARKN), which was formed based on the 11th Meeting of ASOF, and aimed to promote the use of regional knowledge networks to broaden the ASEAN knowledge based on forestry specialised in the issues of forests and climate change (ARKN, 2014). Indonesia allocated funds in its national budget to implement these regimes, and MoFor facilitated some of the ASEAN working group secretariats (like the ASEAN Social Forestry Network). Furthermore, various international actors supported these working groups and engage various actors in Indonesia. For example, there was funding for projects (€9.5 million grant) from the European Commission (EC) in 1995 and a €6.0 million EC-funded project in 2005 for ACB (ACB, 2014). The bureaucracies involved had special plans concerning ASEAN issues as part of their legal planning. However, it seems they used this regime only to look out for their interests in ASEAN positions, and to share knowledge and compete for resources available within the regime (Observation 5, 2014). It is observed that, even though many international actors have successfully promoting the regime, however, domestic non-state actors' shows their low interest to joint.

Asian and Pacific forestry share an information regime: The Asia-Pacific Forestry Commission (APFC) is one of six FAO Regional Forestry Commissions that cover the world's major geographic regions and serves the countries of the region as the largest and most inclusive inter-governmental organisation addressing forests and forestry issues and challenges since 1950 (APFC, 2014). APFC is a regularly published forest outlook statistic that is used by the members, including Indonesia. The ministry responsible for forestry is actively involved in this regime only to profit from sharing information. The other entity in this regime is the Asia Forest Partnership (AFP), which was initiated by the Japanese government and which aimed to promote sustainable forest management in the Asia-Pacific region. It covered several issues, such as (a) good governance and forest law enforcement; (b) the development of a capacity for effective forest management; (c) control of illegal logging; (d) control of forest fires; and (e) the rehabilitation and reforestation of degraded lands (MoFA Japan, 2014). These issues matched those of the Indonesian forestry national plan based on MoFor Regulation 45 of 2011. There are 19 governments, 6 intergovernmental organisations, and 16 civil society organisations involved in AFP (MoFA Japan, 2014), which very much depend on Japanese government support and for which there is no report available that would indicate that the Indonesian government forces bureaucracies to use the AFP regime. Indonesia merely sends its apparatus to this APFC event without producing a detailed follow-up (Observation 1, 2014; Interview 3, 2013).

Tropical rainforest countries' information sharing regime: The eleven tropical rainforest countries are Indonesia, Brazil, Gabon, Costa Rica, Congo, Cameroon, Colombia, Malaysia, Papua New Guinea, Peru and the Democratic Republic of Congo. This association was established by the respective heads of state and government and high-level representatives in 2007. It agreed to cooperate to slow, stop and reverse the loss of forest cover and to promote the rehabilitation of degraded forest lands, forest management and conservation under the Forest Eleven organisation (Forest Eleven, 2009), MoFor is involved in this regime, and there is no obligatory fee for membership in this group (Forest Eleven, 2009). Therefore, there is no report available showing that the Indonesian government may have forced bureaucracies to use this regime, or that it implemented the programmes of the regime. Some non state actors even did not know the regime exist. The relevance is determined by domestic actors using this regime to gain information or to verify information from the other competing tropical forest countries.

4.2. Regimes relevant for agroforestry

Agroforestry is difficult to define, and it involves ambiguous terms (Somarriba, 1992). In the Indonesian context, there is no congruent definition of agroforestry in law or policy (Sahide and Giessen, 2015). Agroforestry is somehow associated with the combination of two aspects in Indonesian legal terms. Firstly, ecological aspect likes *wana tani* (agroforestry or farm forestry) at MoFor Regulation P.91 of 2004 and *hutan sekunder* (secondary forest) at MoFor Regulation 49 of 2011. Secondly, socio-economic aspect likes *hutan kemasyarakatan* (community forest) at MoFor Regulation 37 of 2007 and *hutan rakyat* (people forest) at MoFor Regulation 30 of 2012. It is in line with recent research on Indonesian tropical forest transformation that agroforestry is not only related to the ecological discourse but also income tradeoff between income biodiversity and ecosystem functioning (Steffan-Dewenter et al., 2007) as well as social aspects (Raintree and Warner, 1986). Raintree (1987) argue that agroforestry has a very wide range of ecological and cultural conditions throughout the tropical regions not only in Asia but also in Africa and Latin America. Therefore, we establish independent category for agroforestry regime separately from both in forest regimes and in agriculture regimes. However, the results show that most global forest regimes are also applied to agroforestry regimes, such as (i) biological diversity, (ii) wetlands, (iii) tropical timber trade regime, (iv) climate change, (v) combatting desertification, and (vi) the indigenous rights regime.

4.2.1. Global regimes relevant for agroforestry

Agroforestry certification regime: FSC, LEI, and PEFC not only developed timber certification but also non-timber forest product certification. Indonesia produces many non-timber forest products both in state forest area and non-state forest area, this non timber product has listed in MoFor Regulation P.35 of 2007. FSC developed an FSC NTFP label (Pierce et al., 2003), and LEI developed NTFP certification starting in 2007 (LEI, 2014). Other organisations have certification programmes that could be used for agroforestry. MoA, as the core bureaucracy in agroforestry product development (including NTFPs) finds it difficult to develop agroforestry (including certification), which is grown on forest areas, due to the access restrictions applied to (state) forest areas (Sahide and Giessen, 2015). Even though state actors seem not interesting with this scheme, International and domestic non-governmental organization as well as national agricultural entrepreneur associations is actively introducing this scheme involving state actors, and even some associations have a secretariat space at MoA's building (Observation 1, 2014). Furthermore, another voluntary certification bodies is the SAN/RA (Sustainable agriculture network/Rainforest Alliance), which is the oldest and largest coalition of NGOs striving to improve commodity production in the tropics. Funded by researchers and environmentalists in the 1970s and 1980s (SAN, 2014), it also develops criteria for responsible farm management. NTFP certification is recognised by MoFor as a voluntary certification but neither the MoFor nor the MoA is directly involved in this voluntary mechanism.

4.2.2. Regional regimes relevant for agroforestry

All regional regimes for native forest transformation also apply to agroforestry; see Section 4.1.2.

4.3. Regimes relevant for rubber plantations

4.3.1. Global regimes relevant for rubber plantations

International rubber trade regime: The first International Natural Rubber Agreement (INRA) was reached in 1979, within the framework of the Integrated Programme for Commodities of the UN Conference on Trade and Development (UNCTAD, 2012). This egime aimed to stabilise the prices for natural rubber (NR) on the international market in the medium/long-term markets, as a contribution to a balanced growth of supply and demand (Naturrubber, 1999). Indonesia is facing problems with the low price of rubber (Berita Satu, 2014). There is no implementing international organisation for INRA, since the International Natural Rubber Organisation (INRO) dissolved on 13 October 1999 due to the lack of interest of rubber producing countries in the organisation controlling the rubber market (MoT, 2014a). There were similar disappointing issues with the Association of Natural Rubber Producing Countries (ANRPC), founded in 1970 (MoT, 2014b). Consequently, there is no especially strong global actor within the rubber trade regime, but Indonesia, through the MoT and in coordination with the MoA and the consultation with Indonesian rubber entrepreneur association, is still actively involved in ANRPC. Inherently, international and domestic private actors, particularly those from rubber enterpreneur business, are actively working with state actors on using the regime

4.3.2. Regional regimes relevant for rubber plantations

The Southeast Asian free trade regime and Asia Pacific free trade regimes that apply to native forests and agroforestry also apply to rubber plantations (see Section 4.1.2).

The Southeast-Asian rubber trade regime: In response to a situation where there was no implementing international organisation for the INRA (see Section 4.3.1), the three biggest rubber producing countries in ASEAN (Thailand, Indonesia, and Malaysia) formed a company consortium, the "International Rubber Consortium Limited (IRCo)", under the MoU in 2002 to control SMS (Supply Management Scheme) and AETS (Agreed Export Tonnage Scheme) (IRCo, 2013; MoT, 2014b). IRCo recommended not to expand the rubber production area and recommended only intensification (Republika, 2008). Therefore, the government allocated IDR 78 billion for rubber intensification (Republika, 2014c). Rubber has been chosen as the priority commodity in the Indonesian master plan of development (Ministry of National Plan, 2011) and rubber is also one of the core commodities in the ASEAN Economic community 2015 (MoFor, 2014c). Particular issues about rubber have been raised, such as the low international price (Berita Satu, 2014), the contesting responsible for rubber management in Indonesia, whether rubber in forest areas is administered by MoFor and whether rubber plantations in non-forest areas are administered by the MoA (Sahide and Giessen, 2015). The MoT coordinates with the MoA and represents Indonesian interests in this regime. ASEAN rubber entrepreneurship associations formed a regional organisation called the ASEAN Rubber Business Council (ARBC) for the establishment of a closer rapport through market information exchange, consulting and discussion of common trade practices, monitoring the progress and development of the NR industry in member countries, and fostering closer coordination, cooperation and goodwill between the private sectors and among the member organisations (ARBC, 2008). The other regional actor is the International Tripartite Rubber Organization (ITRO). The three most important producers, namely Indonesia, Malaysia and Thailand, established ITRO in order to manage rubber production, in an effort to maintain an orderly market growth and to guarantee a minimum price to their domestic producers (UNCTAD, 2014).

4.4. Regimes relevant for palm oil plantations

4.4.1. Global regime for palm oil plantations

Palm oil certification regime: Indonesian palm oil has been chosen as the priority commodity in the Indonesian development master plan (Ministry of National Plan, 2011). At least, there are two international palm oil certification bodies working on Indonesia. Firstly, the Roundtable on Sustainable Palm Oil (RSPO) is a global multi-stakeholder initiative on sustainable palm oil, initiated by the WWF (World Wide Fund for Nature) and Migros in 2003, due to palm oil consumers' concern with sustainability. Indonesia and Malaysia, as the biggest palm oil producers, were actively involved in it through their palm oil enterprise associations, ever since the RSPO was formed. It set its headquarters in Malaysia, and then built the RSPO Indonesia Liaison Office (RILO), in 2006 (Sawit Watch, 2007). Members and participants come from many countries that produce or use palm oil. RSPO Certification is a seal of approval certifying that the palm oil used in the product is indeed so produced, and that volumes are traceable (RSPO, 2014). Secondly, the International Sustainability and Carbon Certification (ISCC) is Germany's first certification system for biofuels, including palm oil, since 2010, and the European Commission recognised ISCC as one of the first certification schemes to demonstrate compliance with the EU Renewable Energy Directive's (RED) requirements in 2011 (ISCC, 2014). Several Indonesian big palm oil companies applied for this. For example, Asian Agri, through *PT Inti Indosawit Subur*, received this certificate (Kompas, 2012), and (iii) SAN/RA: this is explained in the agroforestry section (see Section 4.2.1). Based on Observation 4 (2013), Indonesia has enhanced its public authority by creating the ISPO standard (based on MoA Regulation 19 of 2011) to compete with the RSPO and other international, private, voluntary-basis certificates. Due to the nature of this voluntary scheme, Indonesian bureaucracies like the MoA are not involved directly in these schemes. However, the MoA, through the ISPO commission and RSPO, initiated collaboration for a joint study programme under the United Nations Development Programme (ISPO Commission, 2013b; RSPO, 2013). Furthermore, GAPKI (Indonesian palm oil enterprise association) announced that they would formally abandon the RSPO and support for the Indonesian Sustainable Palm Oil (ISPO) in 2011. GAPKI argued that RSPO was more inclined to support consumers' interests, rather than producers' interests, but GAPKI still allows its members to apply for RSPO, or other schemes, due to the voluntary nature of those schemes (Agro Indonesia, 2011; Observation 4, 2013). However, other important private and non-governmental organizations in Indonesia align with other international actors on using and promoting the regime.

4.4.2. Regional regimes relevant for palm oil plantations

There is no specific regime applied on this section. However, *the Southeast Asian free trade*, *the Asia Pacific free trade*, and *the Southeast Asian forest and environmental* regimes apply for the regional palm oil plantations category. For example under *the Southeast Asian forest and environmental* regime, ASEAN formed the ASEAN Vegetable Oils Club, the working groups of which are the MoA's responsibility on palm oil sectors (MoFa, 2014c).

5. Discussion and conclusion

Fig. 6 provides an overview of the international regime complex for tropical rainforest transformation, including its relevance in Indonesia along two analytical dimensions. Our results show a clear distinction between global and regional regimes for all types of forest transformations. This distinction into global and regional regimes will help to understand how Indonesian actors use and create regimes according to the different interest in both arenas, for example, using a regional regime as a buffer against global conservation regimes or strengthening global regimes through regional regimes (similar Giessen et al., unpublished).

These global and regional regimes are supplemented by bilateral agreements. An example is a Letter of Intent signed in 2012 by Indonesia and Norway, in which Norway agreed to contribute up to US$1 billion towards Indonesian REDD+ (including the loss of peatland in Indonesia, which now accounts for more than 80%

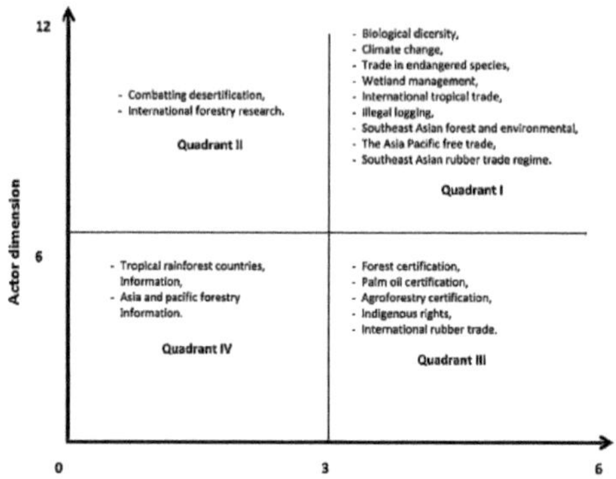

Fig. 6. The regime complex for tropical rainforest transformation of global and regional regimes and its relevance in Indonesia along actor and problem dimensions.

of the country's greenhouse gas emissions) (WRI, 2014 and Satgas REDD, 2014, for details see Brockhaus et al., 2012, 2014; Lesniewska and McDermott (2014); Nurrochmat et al., 2014). Such bilateral regimes, however, for pragmatic reasons were beyond the scope of the analysis.

5.1. Regimes being relevant and being made relevant

Regimes with quadrant I relevance are the regimes on biological diversity, climate change, trade in endangered species, wetland management, international tropical trade, illegal logging, Southeast Asia free trade, Asia Pacific free trade, and Southeast Asian rubber trade. Legality for timber is categorised as an illegal logging regime because legality governance is the tools for combatting illegal logging through a set of domestic legal rules across the forest products supply chain, from forest to consumer. We found that those regimes with quadrant I relevance have a potential impact on Indonesian land use policy. This is in line with hypotheses by Sahide and Giessen (2014) suggesting that such international influences affect rainforest transformation systems. However, the degree of impact is outside the scope of the present study and needs further investigation.

A regime cannot do anything on its own, it only can be used, or it can lead to the formation of new task force bureaucracies or even trigger "super bureaucracies" for the coordination of sectoral bureaucracies. For example, the moratorium policy on logging in natural forest and peatland areas introduced by Presidential Instruction 6 of 2013 is a result of the influence of the climate change regime. This policy is geared politically against the palm oil sector (e.g. palm oil farmer associations and palm oil enterprise associations), in that it reduced palm oil expansion and development in Indonesia (Tempo, 2012; Sawit Indonesia, 2014; Radar Pena, 2014). Furthermore, this supports findings by van Noordwijk et al. (2013) who observed that national interests in retaining global palm oil exports achieved priority over the expectations of the REDD forest rent regime. Some special units or task force bureaucracies that are influenced by international regimes (like the National Council on Climate Change and the REED+ Agency) have now become important bureaucracies involved in land use policy. These bureaucracies have enough power to influence core bureaucracies (like the ministry responsible for forestry and environment) that determine land use transformation policies.

5.2. Regimes being relevant

The results show that the forest certification regime, the agroforestry certification regime, the indigenous rights regime, palm oil certification regime, the international rubber trade regime, the Southeast Asian forest and environmental regime, and the Asia and Pacific forestry information regime are relevant only in the problem dimension.

Those regimes with high problem relevance and no actor backing are urgent but are not on the political agenda. However, these regimes still exist and provide opportunities for future action if domestic actors deem them useful. For example, shared information regimes play important roles in acquiring much information or in verifying information to maintain Indonesian interests in its tropical rainforest transformation. Information is power, especially verified information or pure information, which Indonesian bureaucracies have used in part to enhance their political influence (Simon, 1981; Krott, 1990, 2005; Krott et al., 2013). In this context, this information could be used to oversee the global trends in tropical rainforest transformation. Furthermore, bureaucracies use it to seek advantages in the markets and to balance economic power between tropical forest countries. For example, the Ministry of Agriculture tries to collaborate with voluntary-global of palm oil certification to show that their state certification have similar standard and ready to enter global markets.

Special attention is paid to the global rubber trade regime, which is only relevant in the problem dimension. In contrast, the regional Southeast Asian rubber trade regime is relevant in both dimensions. Indonesian actors profit much more from their involvement in the regional rubber trade regime than in the international rubber trade regime due to the fact that in the regional regime Indonesia and Southeast Asian countries can play their roles in a coalition as the largest rubber-producing countries. This suggests that regional regimes have great potential to be of greater use to domestic actors than global regimes, due to a lower number of veto players and to shared geographical interests in, for example, tropical forestry (similar Bhagwati, 2008; Jürging and Giessen, 2013), a phenomenon which might also apply to biodiversity conservation issues in the region.

5.3. Regimes being made relevant

Regimes that show high relevance only in the actor dimension and low relevance in the problem dimension do not exist frequently. However, our results show that regimes involved in combatting desertification and international forestry research regimes do exist and are relevant only in the actor dimension but not relevant in the problem dimension. Some forest policies have been based on the collaboration of policy research with these regimes' actors. We found that domestic actors use it to back up their arguments for policy development that fit in with domestic political agendas. For example, the REDD+ Agency or the MoFor may form a partnership with the international research community in order to assess particular programmes and/or policies to back up their forest governance. However, Clark et al. (2002) find that CGIAR's research programs that engaging different actors is not harming science, but it need multiple forms of knowledge to avoid the extreme politicization of formal knowledge.

In addition, domestic action arose during the regimes' involvement, due to the specific interests of Indonesian bureaucracies that stood to gain from them, for example, by acquiring more information or verifying information, forming new international coalitions or enhancing existing ones, and even by finding funding sources. The relevance of regimes that have little problem-oriented justification is expected to vanish in the long run. In the short run, however, they have the potential to create substantial political benefits for the actors using them.

5.4. Non-relevant regimes

Information on tropical rainforest countries and Asian and Pacific forestry, as well as international forestry research has been identified as a regime that has no relevance in either dimension. Hence, although it is part of the international regime complex on rainforest transformation, it currently does not address problems relevant to Indonesia, nor is it useful enough for Indonesian key actors to put their own resources behind it.

Both these aspects may change in the future. For example, some forest policies have been based on the collaboration of policy research with these regimes' actors. We found that domestic actors use it to back up their arguments for policy development that fit in with domestic political agendas. Take the REDD+ case as an example; REDD+ Agency or the MoFor may form a partnership with the international research community in order to assess particular programmes and/or policies to back up their forest governance.

5.5. The role for non-state actors

Non-state actors both on domestic and international level is actively on the coalition to promote and use the regime, particularly for transnational regime, or those regime that initiated by private and non-governmental actors like palm oil certification regime and forest certification regime. It is inline with Pattberg (2005) that the collaboration between business and non-profit organizations creating a solid transnational regime. However, the results show that this collaboration is not enough strong to make it relevance for whole actor dimension assessment due to the blocking effort of state actor on reclaiming back their authority, which is inline with Sahide et al., (unpublished). In contrast with the Southeast Asian forest and environment regime and indigenous right regime, non-state actors only have strong consolidation (with state actors) in international and regional level, not on domestic level. It is indicated that state actors still block the area of regime to impose their interest.

Furthermore, this collaboration is created through the pathways of market based interest and direct access from international organizations. It is inline with Bernstein and Cashore (2012) that regimes that can't influence state actors on domestic level through rules and norms, private actors will have change to make it through direct access pathway.

5.6. Regime theory

We reflect regime theories that include regime orientation, non-regimes, and future research ideas, particularly for land use and agricultural regimes. For future research, we observed that the ministerial responsible for trade and industry play important roles as the core bureaucracies involved with the regional regimes (such as ASEAN and APEC). These regimes are associated mostly with trade as their issue for supporting tropical transformations, and are categorised as being relevant in the problem dimension (the first and the third quadrants). Furthermore, they seem to have an important function as buffers to keep environmental issues away from Indonesia. For example, the Ministry of Trade struggles to support the inclusion of Indonesian palm oil and rubber commodities in the Environmental Goods List in the APEC forum. They also seem to have another important buffer function in keeping biodiversity and other conservation issues off the Indonesian agenda, something that is opposed within the Ministry of Environment and Forestry as well as Ministry of Agriculture positions. Furthermore, several environmental or forestry working groups that are add-on regional regimes, like ASEAN, act not only as buffers for regimes, but potentially could also be categorised as non-regimes that are deliberately kept as weak and hollow institutions (Dimitrov et al., 2007).

The method of common indicators for a set of very diverse regimes is successfully tested to measure the relevance, however, it still needs deeper investigation, preferably also for specific regimes instead of regime complexes. For example, climate change regime and regional forest and environmental regime is very high complex to put into one complex as this research delivered. It is recommended to develop more specific indicators for those regimes as one regime complex.

Furthermore, A major shift in structure, priorities and programs due to new government of Indonesia in the end of 2014, will also potentially change the relevance of regime complex. Future studies could be addressed the recent development.

5.7. The urgency for practical policies

This analysis is important material to generally assess the utility of each regime for Indonesian actors as well as international (and regional) actors. This is a general landscape of multiple global and regional land use regimes that can be useful for actors to reconsider their political and practical interests on using the regimes.

Acknowledgements

We greatly appreciate the financial support from the DFG research project on the fragmentation of the international forest regime complex (PAK 813), the Eva Mayr-Stihl Foundation and the Indonesian Government fund (DIKTI scholarship). We are also very grateful to the Editor of the Journal of Land Use Policy and two anonymous reviewers for their insightful and constructive comments.

References

Abbott, K.W., 2011. 2011. The transnational regime complex for climate change. Environ. Plann. C Government Policy 2012 vol. 30, 571–590, http://dx.doi.org/10.1068/c11127

Allison, E.H., Perry, A.L., Badjeck, M.C., Neil Adger, W., Brown, K., Conway, D., Dulvy, N.K., 2009. Vulnerability of national economies to the impacts of climate change on fisheries. Fish Fisheries 10 (2), 173–196.

Arts, B., 2012. Forests policy analysis and theory use: overview and trends. For. Policy Econ. 16 (March), 7–13.

Arts, B., Giessen, L., Visseren-Hamakers, I., 2013. International forest policy and Europe: Four pathways of mutual influence. In: Pülzl, H., et al. (Eds.), European Forest Governance – Issues at stake and the way forward. What Science can tell us 2. EFI, Joensuu, pp. 37–43.

Bernstein, S., Cashore, B., 2012. Complex global governance and domestic policies: four pathways of influence. Int. Affairs 88 (3), 585–604.

Bhagwati, J., 2008. Termites in the World Trading System: How Preferential Agreements Undermine Free Trade. Oxford University Press, New York.

Brockhaus, M., Obidzinski, K., Dermawan, A., Laumonier, Y., Luttrell, C., 2012. An overview of forest and land allocation policies in Indonesia: is the current framework sufficient to meet the needs of REDD+? For. Policy Economics 18, 30–37 http://dx.doi.org/10.1016/j.forpol.2011.09.004

Brockhaus, M., Di Gregorio, M., Mardiah, S., 2014. Governing the design of national REDD+: an analysis of the power of agency. For. Policy Econ. 49, 23–33 http://dx.doi.org/10.1016/j.forpol.2013.07.003

Budidarsono, S., Susanti, A., Zoomers, A., 2013. Oil Palm Plantations in Indonesia: The Implications for Migration, Settlement/Resettlement and Local Economic Development.

Budiman, A., Penot, E., 1997. Smallholder rubber agroforestry in Indonesia. In International Rubber Conference, IRC. Kuala Lumpur.

Burns, S.L., Giessen, L., 2014. Identifying the main actors and their positions on international forest policy issues in Argentina. BOSQUE 35 (2), 163–173, http://dx.doi.org/10.4067/S0717-92002014000200004

Clark, W.C., Tomich, T.P., van Noordwijk, M., Guston, D., Catacutan, D., Dickson, N.M., McNie, E., 2011. Boundary work for sustainable development: natural resource management at the Consultative Group on International Agricultural Research (CGIAR). Proc. Natl. Acad. Sci., 200900231.

Dauvergne, P., 1998. The political economy of Indonesia's 1997 forest fires. Aust. J. Int. Affairs Vol. 52 (No.1), 13–17.

Derkyi, M., Ros-Tonen, M.A., Kyereh, B., Dietz, T., 2013. Emerging forest regimes and livelihoods in the Tano Offin Forest Reserve, Ghana: implications for social safeguards. For. Policy Econ. Vol. 32 (July), 49–56.

Dequin, H., 1975. Land resources in Indonesia. Geoforum 6 (3), 257–260.

Dhiaulhaq, A., Gritten, D., De Bruyn, T., Yasmi, Y., Zazali, A., Silalahi, M., April 2014. Transforming conflict in plantations through mediation: lessons and experiences from Sumatera, Indonesia. For. Policy Econ. Vol. 41, 22–30.

Dimitrov, R.S., Sprinz, D.F., DiGiusto, G.M., Kelle, A., 2007. International Nonregimes: a research agenda 1. Int. Stud. Rev. 9 (2), 230–258.

Edwards, P., Kleinschmit, D., August 2013. Towards a European forest policy – conflicting courses. For. Policy Econ. Vol. 33, 87–93.

Edwards, P., Giessen, L., 2014. Global forest governance – discussing legal scholarship from political science perspectives. For. Policy Econ. 38 (2014), 30–31 http://dx.doi.org/10.1016/j.forpol.2013.08.006

Feintrenie, L., Levang, P., 2009. Sumatra's rubber agroforests: advent, rise and fall of a sustainable cropping system. Small-Scale For. 8 (3), 323–335.

Feintrenie, L., Chong, W.K., Levang, P., 2010. Why do farmers prefer oil palm? Lessons learnt from Bungo district, Indonesia. Small-Scale For. 9 (3), 379–396.

Fitzherbert, E.B., Struebig, M.J., Morel, A., Danielsen, F., Brühl, C.A., Donald, P.F., Phalan, B., 2008. How will oil palm expansion affect biodiversity? Trends Ecol. & Evol. 23 (10), 538–545 http://dx.doi.org/10.1016/j.tree.2008.06.012

Hofmann, F., 2002. Globale Waldpolitik in Deutschland-Eine Untersuchung über die Wirkung internationaler Regime in föderalen Strukturen. In: PhD Thesis. Albert-Ludwigs-Universität, Freiburg, 311 p.

Hogl, K., Nordbeck, R., Kvarda, E., 2009. When international impulses hit home: the role of domestic policy subsystem configurations in explaining different types of sustainability strategies. For. Policy Econ. 11 (5–6), 357–364.

Humphreys, D., 2006. Logjam – Deforestation and the Crisis of Global Governance. Earthscan, London.

Jürging, D.J.J., Giessen, L., 2013. Ein Rechtsverbindliches Abkommen über die Wälder in Europa: stand und Perspektiven aus rechts-und umweltpolitikwissenschaftlicher Sicht. Natur und Recht 35 (5), 317–323.

Giessen, Bezerra, Sindt, Jetschke, unpublished. The Rational Design of Regional Forest Regimes: Comparing Amazonian, Central African and Pan-European Forest Cooperation.

Giessen, L., 2012. Temporary governance and persistent government – Rural policy integration in pilot and mainstream funding programmes. In: Hogl, K., Kvarda, E., Nordbeck, R., Pregernig, M. (Eds.), Environmental Governance: The Challenge of Legitimacy and Effectiveness. Edward Elgar, pp. 155–177.

Giessen, L., 2013a. Reviewing the main characteristics of the international forest regime complex and partial explanation for its fragmentation. Int. For. Rev. 15 (1), 60–70.

Giessen, L., 2013b. Fragmentierung als Schlüsselfaktor des internationalen Waldregimes: von einem mono- zu einem multi-disziplinären methodischen Rahmen für eine vertiefte Forstpolitikforschung. Allgemeine Forst- und Jagdzeitung 184 (3–4), 47–57.

Giessen, L., Krott, M., 2009. Forestry Joining Integrated Programmes? A question of willingness, ability and opportunities. Allgemeine Forst- und Jagdzeitung 180 (5–6), 94–100.

Giessen, L., Krott, M., Möllmann, T., 2014. Increasing representation of states by utilitarian as compared to environmental bureaucracies in international forest and forest-environmental policy negotiations. For. Policy Econ. 38, 97–104 http://dx.doi.org/10.1016/j.forpol.2013.08.008

Humphreys, D., 2009. Discourse as ideology: Neoliberalism and the limits of international forest policy. For. Policy Econ. 11 (5), 319–325.

Hovi, J., Sprinz, D., Underdal, A., 2003. Regime Effectiveness and the Oslo- Potsdam Solution: A Rejoinder to Oran Young. Global Environmental Politics 3:3, August 2003 © 2003 by the Massachusetts Institute of Technology.

Hsieh, H.F., Shannon, S.E., 2005. Three approaches to qualitative content analysis. Qual. Health Res. 15, 1277–1288.

Keohane, R.O., Victor, D.G., 2011. The regime complex for climate change. Perspect. Politics 9 (1), 7–23.

Klooster, D., 2010. Standardizing sustainable development? The forest stewardship council's plantation policy review process as neoliberal environmental governance. Geoforum 41 (1), 112–117, http://dx.doi.org/10.1016/j.geoforum.2009.02.006

Krasner, S.D., 1982. Structural causes and regime consequences: regimes as intervening variables. Int. Organ. 36 (2), 185–205.

Krasner, S.D., 1983. International Regimes. Cornell University Press, Ithaca, pp. 372pp–$9.

Krott, M., 1990. Öffentliche Verwaltung im Umweltschutz. Ergebnisse einer behördenorientierten Policy-Analyse am Beispiel Waldschutz, W. Braumüller Verlag, Wien.

Krott, M., 2005. Forest Policy Analysis. Springer, Dordrecht, 323 pp.

Krott, M., Hasanagas, N.D., 2006. Measuring bridges between sectors: causative evaluation of cross-sectorality. For. Policy Econ. 8 (5), 555–563.

Krott, M., Bader, A., Schusser, C., Devkota, R., Maryudi, A., Giessen, L., Aurenhammer, H., 2013. Actor-centred power: the driving force in decentralised community based forest governance. For. Policy Econ. 49, 34–42.

Lawrence, D.C., 1996. Trade-offs between rubber production and maintenance of diversity: the structure of rubber gardens in West Kalimantan, Indonesia Agrofor. Syst. 34 (1), 83–100.

Lesniewska, F., McDermott, C.L., 2014. FLEGT VPAs: Laying a pathway to sustainability via legality lessons from Ghana and Indonesia. Forest Policy and Economics. 48, 16–23. http://dx.doi.org/10.1016/j.forpol.2014.01.005

Lindstad, B.H., Solberg, B., 2010. Challenges in determining effects of international policy processes: forest protection in Norway as a case. For. Policy Econ. 12 (2010), 489–496.

Lindstad, B.H., 2014. Changes in Norwegian forest policies: what are the effects of the international forest regime complex? Scand. J. For. Res., http://dx.doi.org/10.1080/02827581.2014.1000478

McCarthy, J., Zen, Z., 2010. Regulating the oil palm boom: assessing the effectiveness of environmental governance approaches to agro-industrial pollution in Indonesia. Law Policy 32 (1), 153–179.

Measey, M., 2010. Indonesia: a vulnerable country in the face of climate change. Global Majority E J. 1 (1), 31–45.

McDermott, C.L., Cashore, B., Kanowski, P., 2010. Global environmental forest policies: An international comparison. Earthscan, London, Washington, DC.

McDermott, C.L., 2012. REDDuced: From sustainability to legality to units of carbon- The search for common interests in international forest governance. Environ. Sci. Policy, http://dx.doi.org/10.1016/j.envsci.2012.08.012

Miyamoto, M., 2006. Forest conversion to rubber around Sumatran villages in Indonesia: comparing the impacts of road construction, transmigration projects and population. For. Polic Econ. 9 (1), 1–12.

Miyamoto, M., Parid, M.M., Aini, Z.N., Michinaka, T., 2014. Proximate and underlying causes of forest cover change in Peninsular Malaysia. For. Policy Econ. Vol. 44 (July 2014), 18–25.

MoFor, 2012. Indonesia Forest Management Policy Reform in Addresing Climate Change, Ministry of Forestry.

Mosley, L., 2013. Introduction. just talk to people? interviews in contemporary political science. In: Mosley, L. (Ed.), Interview Research in Political Science. Cornell University Press, p. 2013.

Munadi, E., 2007. Penurunan Pajak Ekspor dan Dampaknya terhadap Ekspor Minyak Kelapa Sawit Indonesia ke India (Pendekatan Error Correction Model). Informatika Pertanian 16 (2), 1020–1036.

Neuman, W.L., 2005. Social research methods. In: Quantitative and Qualitative Approaches, 6th ed. UK Allyn and Bacon, London.

Neyrma, M., Neyra-Cabatac, Juan M. Pulhin, Daylinda B. Cabanilla, 2012. Indigenous agroforestry in a changing context: the case of the Erumanen ne Menuvu in Southern Philippines. For. Policy Econ., Vol. 22, September 2012, 18–27.

Nurrochmat, D.R, Dharmawan, A.H, Obidzinski, K., Dermawan, A., Erbaugh, J.T., 2014. Contesting national and international forest regimes: Case of timber legality certification for community forests in Central Java, Indonesia. For. Policy Econ., in Press, Corrected Proof, Available online 30 September 2014, http://dx.doi.org/10.1016/j.forpol.2014.09.008

Orsini, A., Morin, J.F., Young, O., 2013. Regime complexes: a buzz, a boom, or a boost for global governance? Global Governance Rev. Multilateralism Int. Organ. 19 (1), 27–39.

Overdevest, C., Zeitlin, J., 2014. Constructing a transnational timber legality assurance regime: architecture, accomplishments, challenges. For. Policy Econ. Vol. 48 (November 2014), 6–15.

Pattberg, P., 2005. The institutionalization of private governance: how business and nonprofit organizations agree on transnational rules. Governance 18 (4), 589–610.

Purnomo, A., Katili-Niode, A., Melisa, E., Helmy, F., Sukadri, D., Sitorus, S., 2013. Evolution of Indonesia's climate change policy, from Bali to Durban. National Council on Climate Change, Jakarta.

Peters, B.G., 2010. The Politics of Bureaucracy – An Introduction to Comparative Public Administration, 6th ed. Routledge, Oxon.

Rahman, Md. S., Giessen, L., 2014. Mapping international forest-related issues and main actors' positions in Bangladesh. Int. For. Rev. vol. 6 (6), 1–16.

Raintree, J.B., Warner, K., 1986. Agroforestry pathways for the intensification of shifting cultivation. Agrofor. Syst. 4 (1), 39–54.

Raintree, J.B., 1987. The state of the art of agroforestry diagnosis and design. Agrofor. Syst. 5 (3), 219–250.

Raustiala, K., Victor, David G., 2004. The regime complex for plant genetic resources. Int. Organ. 55, 277–309.

Rayner, J., Howlett, M., Wilson, J., Cashore, B., Hoberg, G., 2001. Privileging the sub-sector: critical sub-sectors and sectoral relationships in forest policy-making. For. Policy Econ. 2 (3), 319–332.

Rayner, J., Humphreys, D., Perron Welch, P., Frederic, Prabhu, Ravi, Verkooijen, P., 2010. Introduction. In: Rayner, J., Buck, A., Katila, P. (Eds.), Embracing Complexity: Meeting the Challenges of International Forest Governance, vol. 28. IUFRO World Series, Vienna, pp. 9–18.

Sadath, N., Krott, M., 2012. Identifying policy change-analytical program analysis: an example of two decades of forest policy in Bangladesh. For. Policy Econ. 25, 93–99.

Sahide, M.A.K., Giessen, L., 2015. The fragmented land use administration in Indonesia – analysing bureaucratic responsibilities influencing tropical rainforest transformation systems. Land Use Policy 43 (2005), 96–110, http://dx.doi.org/10.1016/j.landusepol.2014.11.005

Sahide, M.A.K., Burns, S., Wibowo, A., Nurrochmat, D.R., Giessen, L., (unpublished). The changing governance of agrarian certification: from voluntary private governance to mandatory state labelling for sustainability in Indonesia's palm oil sector.

Sadath, N., Kleinschmit, D., Giessen, L., 2013. Framing the tiger – a biodiversity concern in national and international media reporting. For. Policy Econ. 36, 37–41.

Simon, H.A., 1981. Entscheidungsverhalten in Organisation. Deutsche Übersetzung, 3. Landsberg, Auflage.

Smouts, M.-C., 2008. The issue of an International Forest Regime. Int. For. Rev. 10 (3), 429–432 http://www.bioone.org/doi/full/10.1505/ifor.10.3.429

Somarriba, E., 1992. Revisiting the past: an essay on agroforestry definition. Agrofor. Syst. 19 (3), 233–240.

Sprinz, D., 2005. Regime Effectiveness: The Next Wave of Research. Potsdam Institute for Climate Impact Research & The University of Potsdam.

Steffan-Dewenter, I., Kessler, M., Barkmann, J., Bos, M.M., Buchori, D., Erasmi, S., Forest areaust, H., Gerold, G., Glenk, Gardstein, S.R., Guhardja, E., Hartereld, M., Hertel, D., Höhn, P., Kappas, M., Köhler, S., Leuscher, C., Maertens, M., Marggraf, R., Migge-Kleian, S., Mogea, J., Pitopang, R., Schaefer m Schwarze, S., Sporn, S.G., Steingrebe, A., Tjitrosoedirdjo, S.S., Tjitrosoemito, S., Twele, A., Weber, R., Woltmann, L., Zeller, M., Tscharntke, T., 2007. Tradeoffs between income, biodiversity, and ecosystem functioning during tropical rainforest conversion and agroforestry intensification. Proc. Natl Acad Sci. 104 (12), 4973–4978.

Van Evera, S., 1997. Guide to Methods for Students of Political Science. Cornell University Press.

van Noordwijk, M., Agus, F., Dewi, S., Purnomo, H., 2013. Reducing emissions from land use in Indonesia: motivation, policy instruments and expected funding streams. Mitigation Adapt. Strategies Global Change, 1–16, http://dx.doi.org/10.1007/s11027-013-9502-y

Wibowo, A., Giessen, L., 2012. Identifying international forest-related issues in Indonesia based on the statements in public and expert deliberations. Int. J. Social For., in press.

Young, O.R., 2001. Inferences and Indices: Evaluating the Effectiveness of International Environmental Regimes. Global Environmental Politic 1:1, 99–121.

Empirical references

ACB, 2014. http://www.aseanbiodiversity.org/index.php?option=com_content&view=article&id=69&Itemid=79&current=1 (accessed 11.03.14.).

Agro Indonesia, 2011. RSPO untuk Kepentingan Siapa?, http://agroindonesia.co.id/2011/10/25/rspo-untuk-kepentingan-siapa/ (accessed 03.04.14.).

AMAN, 2011. Indigenous World 2011, www.aman.or.id/?file_id=6 (accessed 30.04.14.).

Antara, 2013. Manfaat APEC untuk Indonesia, http://www.antaranews.com/berita/398251/manfaat-apec-untuk-indonesia (accessed 04.05.14.).

Antara, 2013. Hutan Adat Terancam, Presiden SBY Diminta Turun Langsung, http://nasional.news.viva.co.id/news/read/498792-hutan-adat-terancam-presiden-sby-diminta-turun-langsung (accessed 25.11.14.).

Anwar, C., 2006. Perkembangan Pasar dan Prospek Agribisnis Karet di Indonesia. Lokakarya Budidaya Tanaman Karet, 1–19.

APFC, 2014. Welcome to the Asia–Pacific Forestry Commission, http://www.fao.org/forestry/33587/en/ (accessed 11.03.14.).

ARBC, 2008. About us, http://www.aseanrubber.org/arbc/index.php?option=com_content&view=article&id=55&Itemid=87 (accessed 26.11.14.).

ASEAN, 2014. AEAN Haze action online, http://haze.asean.org/?page_id=185 (accessed 11.03.14.).

ASFN, 2014. http://www.asfnsec.org/general/faq (accessed 04.05.14.).

Berita Satu, 2014. Harga Karet Turun, DPR Minta Pemerintah Bangun Industri Pengolahan Domestik, http://www.beritasatu.com/industri-perdagangan/170314-harga-karet-turun-dpr-minta-pemerintah-bangun-industri-pengolahan-domestik.html (accessed 26.11.14.).

Boer, R., 2001. Economic assessment of mitigation options for enhancing and maintaining carbon sink capacity in Indonesia. Mitigation and Adaptation Strategies for Global Change, 6 (3–4), 257–290.

CBD, 2015a. http://www.cbd.int/history/ (accessed 07.04.15.).

CBD, 2015b. http://www.cbd.int/cooperation/ (accessed 07.04.15.).

CITES, 2014. What is CITES? http://www.cites.org/eng/disc/what.php (accessed 26.03.14.).

CIP, 2013. https://www.climateinvestmentfunds.org/cif/home (accessed 11.03.14.).

CIP, 2014. https://www.climateinvestmentfunds.org/cif/node/5 (accessed on 11.03.14.).

Constitutional Court Decision 35 of 2012. Keputusan Mahkamah Konstitusi Nomor 35/PUU-X2012.

Detik, 2014. Pemerintah Siapkan Rp 5 Miliar untuk Subsidi Pembuatan Sertifikat Kayu. http://finance.detik.com/read/2014/03/14/121641/2525741/4/pemerintah-siapkan-rp-5-miliar-untuk-subsidi-pembuatan-sertifikat-kayu (accessed on 11.03.14.).

EU FLEGT Facility, 2015. http://www.euflegt.efi.int/indonesia (accessed 07.04.15.).

Evans, K., 2014. REDD+ on the ground: For one initiative in Indonesia, politics in the peatlands. http://blog.cifor.org/25554/for-redd-initiative-in-central-kalimantan-politics-in-the-peatlands#.VSO2qFwxGRs (accessed 07.04.15.).

FAO, 2014. Sustainable agriculture and rural development, http://www.fao.org/docrep/u8480e/u8480e0l.htm (accessed 12.03.14.).

FORCLIME, 2014. http://www.forclime.org/en/ (accesed 11.03.14.).

Forest Legality Alliance, 2014. Forest Legality Basics, http://risk.forestlegality.org/forest-legality-basics (accessed 27.03.14.).

Forest Eleven (F-11), 2009. http://www.forest-eleven.dephut.go.id/index.php?kode=6&pre=0 (accessed 02.04.14)

GAPKINDO, 2014a. Luas perkebunan karet Indonesia 2008–2013, http://www.gapkindo.org/index.php/id/component/content/article/1-artikel/157-luas-perkebunan-karet-id.html (accessed 14.04.14.).

GAPKINDO, 2014b. Produksi karet alam Indonesia 2008–2013. http://www.gapkindo.org/index.php/id/component/content/article/1-artikel/153-perkebunan-karet-alam-id.html (accessed 14.04.14.).

Government Regulation 8 of 1999 on using Endangered Species of Wild Fauna and Flora. Peraturan Pemerintah Nomor 8 tahun 1999 tentang Pemanfaatan Jenis Tumbuhan dan Satwa Liar.

Gouyon Anne, 2003. Eco-Certification as an Incentive to Conserve Biodiversity in Rubber Smallholder Agroforestry Systems: A Preliminary Study. World Agroforestry Centre (ICRAF), http://www.worldagroforestrycentre.org/Sea/Publications/files/workingpaper/WP0003-04.PDF (accessed 12.03.14.).

ICTSD, 2014. http://ictsd.org/i/publications/170659/#sthash.BmmKTfDO.dpuf (accessed 12.03.14.).

IFCC, 2014. http://ifcc-ksk.org/ifcc/index.php/en (accessed 12.03.14.).

IFOAM, 2014. http://www.ifoam.org/en/about-us-1 (accessed 12.03.14.).

Interview 1, 2014. Interview with high-level staff of Association of Indonesia Forest Concession Holders (APHI), interviewing the trend of International Forest Policy influencing Indonesian Forest Productivity in the APHI office. Jakarta, Indonesia. 19.12.2013.

Interview 2, 2014. Separate interview with officer of MoFor who was implementing one of the ITTO projects in Indonesia, and NGO officer who was collaborating with MoFor implementing ITTO in Indonesia. Jakarta, Indonesia. 15.01.2014 and Bogor, Indonesia 17.01.2014.

Interview 3, 2013. Separate interview with 2 Officers for Expert Staff of Ministry of Forestry on different issues (climate change and indigenous people). Jakarta Indonesia, 02.12.2013 and 17.12.2013.

Interview 4, 2013. Officer for National Council on Climate Change, 2 staff (deputy) interviewed separately in the Office, the staff is usually representing Indonesia on Climate Change issues in International Forum like UNFCC, and the staff is also mediating inter-ministries meeting on Climate Change issues to set Indonesian position. Jakarta, Indonesia, 07.02.2014 and 19.01.2014.

Interview 5, 2014. Member of PEFC in Indonesia, member from academic/university representative, Bogor. Interview about the history of PEFC in Indonesia and its progress. Bogor, Indonesia. 22.01.2014.

Interview 6, 2014. Separate interview with ex-officer (but still the member) of LEI in Indonesia, and the participant member of LEI from NGO representative. Interview about LEI existence in Indonesia. Jakarta, Indonesia. 22.01.2014 and Bogor, Indonesia 17.01.2014.

Interview 7, 2014. Separate interview with Officer responsible for SVLK implementation in Indonesia from Ministry of Forestry and Ministry of Trade. Interview about the progress and the challenge of SVLK implementation. Jakarta, Indonesia, 12.02.2014 and 11.02.2014.

ISCC, 2014. ISCC offers solutions for different markets, http://www.iscc-system.org/en/iscc-system/about-iscc/ (accesed 03.04.14.).

Iskandar U, 1999. Dialog Kehutanan dalam Wacana Global.Bigraf Publishing. Jakarta

ISPO, 2014. http://www.ispo-org.or.id/index.php?option=com_content&view=article&id=52&Itemid=217&lang=en (accessed 03.04.14.).

ISPO Commission, 2013a. Indonesian Palm Oil Statistics 2012.

ISPO Commission, 2013b. http://www.ispo-org.or.id/index.php?option=com_content&view=article&id=93&Itemid=271&lang=en (accessed 25.06.14.).

Joshi, L., Wibawa, G., Vincent, G., Boutin, D., Akiefnawati, R., Manurung, G., Van Noordwijk, M., 2001. Wanatani kompleks berbasis karet: tantangan untuk pengembangan (rubber-based complex agroforestry systems: a challenge for development.

Kompas, 2012. Asian Agri Terima Sertifikasi ISCC, http://properti.kompas.com/read/2012/04/05/16344087/Asian.Agri.Terima.Sertifikasi.ISCC (accessed 11.03.14.).

Law 5 of 1990 on Biodiversity Conservation and its Ecosystem. UU Nomor tahun 1990 tentang Konservasi Sumber Daya Alam Hayati Dan Ekosistemnya.

Law 10 of 2013 on approving Rotterdam convention on the prior informed consent procedure for certain hazardous chemicals and pesticides in international trade. UU Nomor 10 tahun 2013 tentang Pengesahan konvensi Rotterdam tentang prosedur persetujuan atas dasar informasi awal untuk bahan kimia dan pestisida berbahaya tertentu dalam perdagangan internasional.

Law 11 of 2013 on approving Nagoya protocol on access to genetic resources and the fair and equitable sharing of benefits arising from their utilization to the convention on biological diversity. UU Nomor 11 tahun 2013 tentang Pengesahan protokol nagoya tentang akses pada sumber daya genetik dan pembagian keuntungan yang adil dan seimbang yang timbul dari pemanfaatannya atas konvensi keanekaragaman hayati.

Law 38 of 2008 on approving Charter of ASEAN. Undang-Undang Nomor 38 Tahun 2008 tentang Pengesahan Charter of ASEAN.

LEI, 2014. Standar Sertifikasi Hasil Hutan Bukan Kayu, http://www.lei.or.id/news/1218/standar-sertifikasi-hasil-hutan-bukan-kayu (accessed 15.04.14.).

Letter of Intent between Norway and Indonesia http://www.unorcid.org/upload/doc_lib/Norway-Indonesia-LoI.pdf (accessed 15.04.14.).

Manurung, T., 2000. Pembangunan Perkebunan Kelapa Sawit Di Indonesia: Ancaman Terhadap Hutan Alam. Warta Fahutan Online, 1–6.

MoA Regulation 19 of 2011.Guideline on Indonesian Sustainable Palm Oil (ISPO).Peraturan Menteri Kehutanan Nomor 19 Tahun 2011tentang Pedoman Perkebunan Kelapa Sawit Berkelanjutan Indonesia (Indonesian sustainable palm oil/ISPO).

MoA, 2007. Prospek dan Arah Pengembangan Agribisnis Kelapa Sawit.Edisi Kedua.

MoA, 2012. http://www.deptan.go.id/infoeksekutif/bun/BUN-asem2012/Areal-Karet.pdf (accessed 03.06.13.).

MoFA Japan, 2014. Asia Forest Partnership (AFP), http://www.mofa.go.jp/policy/economy/fishery/afp/general0211.html (accessed 02.04.14.).

MoFA, 2014a. Keanggotaan Indonesia pada Organisasi Internasional, http://kemlu.go.id/Documents/Keanggotaan_Indonesia_pada_OI.pdf (accessed 26.03.14.).

MoFa, 2014b. http://www.kemlu.go.id/Pages/AboutUs.aspx?IDP=1 (accessed 04.05.14.).

MoFa, 2014c. Report on Meeting-21 Joint Commitee on ASEAN Cooperation in Agriculture and Forest Promotion Scheme.

MoFa, 2013. Rencana Umum Pengadaan Per Satuan Kerja Kementerian Luar Negeritahun Anggaran 2014.

MoE, 2011. Dana Alokasi Khusus Bidang Lingkungan Hidup Tahun Anggaran 2012, http://www.menlh.go.id/dana-alokasi-khusus-bidang-lingkungan-hidup-tahun-anggaran-2012/#sthash.KbYSQleC.dpuf (accessed 02.05.14.).

MoE, 2012. Menyongsong Ratifikasi Protokol Nagoya, http://www.menlh.go.id/menyongsong-ratifikasi-protokol-nagoya/ (accesed 02.05.14.).

MoE, 2013. Peluang dan Tantangan Protokol Nagoya di Indonesia, http://www.menlh.go.id/peluang-dan-tantangan-protokol-nagoya-bagi-indonesia/ (accesed 26.03.14.).

MoFor, 2012. Indonesia Forest Management Policy Reform in Addresing Climate Change, Bonn.

MoFor, 2004a. Siaran pers No: S.548/II/PIK-1/2004, http://www.dephut.go.id/index.php/news/details/1671 (accesed 26.03.14.).

MoFor, 2004. Peningkatan Peran sektor Kehutanan dalam ASEAN Economic Community 2015. Policy Brief vol. 8. No. 1, Tahun 2014.

MoFor, 2013. Forestry Statistic of Indonesia 2012.

MoFor, 2005. Hutan Rakyat Indonesia Sangat Prospektif untuk Industri Kehutanan.

MoFor, 2014d. Mengenal Indonesia.forest climate alliance (IFCA). http://www.dephut.go.id/uploads/INFORMASI/LITBANG/IFCA/Pengantar.htm (accessed 03.04.14.).

MoFor Degree 104 of 2003 on designating Directorate General for nature protection and conservation as the Management Authority of CITES in Indonesia. Keputusan Menteri Kehutanan Nomor 104/Kpts-II/2003 tentang Penunjukan Direktur Jenderal Perlindungan Hutan dan Konservasi Alam sebagai Otorita Pengelola (Management Authority) CITES di Indonesia.

MoFor Degree 3 of 2004 on the guideline of national land rehabilitation programm. Keputusan Menteri Kehutanan No. P.03/Menhut-V/2004 tentang Pedoman dan Petunjuk Pelaksanaan Penyelenggaraan Gerakan Nasional Rehabilitasi Hutan dan Lahan.

MoFor Degree 390 of 2003 on the Partnership's guideline on Nature Conservation. Keputusan Menteri Kehutanan Nomor: 390/Kpts-II/2003 tentang Tata Cara Kerjasama di Bidang Konservasi Sumber Daya Alam Hayatidan Ekosistemnya.

MoFor Regulation 42 of 2013 on the third revision of MoFor Regulation 38 of 2009 on the standard and guideline of Performance Assessment of Sustainable Production Forest Management and Timber Legality Verification License Holder or in Title Forest. Peraturan Menteri Kehutanan Republik Indonesia Nomor: P.42/Menhut-II/2013 Tentang Perubahan Ketiga Atas Peraturan Menteri Kehutanan Nomor P.38/MENHUT-II/2009 Tentang Standar dan Pedoman Penilaian Kinerja Pengelolaan Hutan Produksi Lestari Dan Verifikasi Legalitas Kayu Pada Pemegang Izin Atau Pada Hutan Hak.

MoFor Regulation 49 of 2011.National Forestry Planning (RKTN) Period 2011–2030, http://dephut.net/uploads/files/DitRenHut_RKTN_2011.pdf (accessed 22.11.14.).

MoFor Regulation51 of 2010. Strategic Plan of MoFor 2010–2014. Peraturan Menteri Kehutanan Nomor: P. 51/Menhut-

/2010 Tentang Rencana Strategis (Renstra) Kementerian Kehutanan Tahun 2010–2014.

MoFor Regulation 38 of 2009 on Satndard and the Guideline of Assesment of Forest Production Management and Timber Legality for Licence Holder of Title Forest. Peraturan Menteri Kehutanan No.38/Menhut-II/2009 tentang Standard dan Pedoman Penilaian Kinerja Pengelolaan Hutan Produksi Lestari dan Verifikasi Legalitas Kayu Pada Pemegang Izin Atau Pada Hutan Hak.

MoFor Regulation 14 of 2004 on the Procedure of Afforestation and Reforestation (A/R) on Clean Development Mechanism. Peraturan Menteri Kehutanan Nomor P.14/Menhut-II/2004 tentang Tata Cara Aforestasid Reforestasi dalam Kerangka Mekanisme Pembangunan Bersih.

MoFor Regulation 37 of 2007 on Community Forestry. Peraturan Menteri Kehutanan P.37/Menhut-II/2007tentang Hutan Kemasyarakatan.

MoFor Regulation P.91 of 2014 on the Administration of Non Timber Forest Products from State Forest. Peraturan Menteri Kehutanan Nomor: P.91/Menhut-II/2014 tentang Penatausahaan Hasil Hutan Bukan Kayu Yang Berasal Dari Hutan Negara.

MoFor Regulation30 of 2012 on the forest Products Administration from Title Forest.Peraturan Menteri Kehutanan P.30/Menhut-II/2012 tentang 2012 tentang Penatausahaan Hasil Hutan Yang Berasal dari Hutan Hak.

Mogabay Indonesia, 2013. Akhirnya, Presiden Tunjuk Heru Prasetyo Pimpin Badan REDD+ http://www.mongabay.co.id/tag/lembaga-redd/ (accessed 31.03.14.).

Ministry of National Planning, 2014. National Action Plan on Climate Change Adaptation.

Ministry of National Plan, 2011. The Master Plan for Acceleration and Expansion of Indonesia Economic Development (MP3EI).

MoI. Pemerintah Anggarkan Rp 315 Miliar Kembangkan Industri Agro. http://agro.kemenperin.go.id/60-Pemerintah-Anggarkan-Rp-315-Miliar-Kembangkan-Industri-Agro (accessed 26.11.14.).

MoSA, 2014. http://www.kemsos.go.id/modules.php?name=News&file=article&sid=1377 (Accessed 30.04.14.).

MoT, 2014a. International Rubber Consortium Limited (IRCo).

MoT, 2014b. Association of Natural Rubber Producing Countries (ANRPC).

MoT Regulation 64 of 2012 on Conditions of Forestry Industry Products Export. Peraturan Menteri Perdagangan Republik Indonesia Nomor 64/M-DAG/PER/10/2012 tentang Ketentuan Ekspor Produk Industri Kehutanan.

Murniningtyas, S., Justianto, A., Pribadi, A., Ida, F., Didi, I., Exploitasia, I., Martana, K., Rufi'ie, Anwar, S., Rahardja, T., Yanuariadi, T., Septiani, Y., 2007. Instrumen Kehutanan Global. Ministry of Forestry.

NRDCF, 2014. http://nrdcf.com/13-14-mei-2014 (accessed 24.11.14.).

Natuurrubber, 2009. Termination of the International Natural Rubber Agreement. Natuurrubber 16–4th quarter 1999.

NG Indonesia, 2011. Indonesia Masuk Daftar Pengelolaan Hutan Terburuk, http://m.nationalgeographic.co.id/berita/2011/11/indonesia-masuk-daftar-pengelolaan-hutan-terburuk (accessed 28.03.14.).

Observation 1, 2014. Observation on the activities of the Centre of Foreign Cooperation oF MoFor and MoA. How this bureaucracy unit coordinating International Forest Regime to other unit bureaucracies under MoFor and MoA or other Ministries, during Field Work I and Field Work IIin Jakarta, Indonesia. 2014

Observation 2, 2013. CIFOR and ICRAF activities in Indonesia. Observing consultancy of CIFOR's and ICRAF'sactivities in Indonesia, particularly on the field of research arena, how they conduct research, and how they produce policy recommendation, as well as CIFOR relation to other key stakeholder in Indonesia, during Field Work I and Field Work II in Bogor and Jakarta, Indonesia. 2014

Observation 3, 2013. Observing contesting issues among key stakeholders on indigenous issues such as AMAN (the most important civil society organisation and well known on Indigenous issues in Indonesia and also other NGOs, Ministry responsible for forestry) as well as the Indonesian Communication Forum on Community Forestry (FKKM), during Field Work I and Field Work II in Bogor and Jakarta, Indonesia. 2014

Observation 4, 2013. Observing contesting issues about mandatory and voluntary schemes of Indonesian palm oil in Indonesia through ISPO and RSPO Indonesia Office, during Field Work I, Jakarta. Indonesia. 2014.

Observation 5, 2013. Observing all active ASEAN working groups on forestry and environmental issues under Ministry of Forestry and Ministry of Environment, as well as the ASEAN activities for trade issues during Field Work I and Field Work II (October–November 2014) in Jakarta.

Observation 6, 2014. Observing the informal discussion about interest behind the integration of Ministry of Forestry and Ministry of Environment become Ministry of Environment and Forestry, as well as its influence of international forest and environment implementation, during Field Work II in Bogor and Jakarta, Indonesia.

PEFC, 2014b. http://www.pefc.org/projects/knowledge/smallholder-group-certification (accessed on 29.04.14.).

PEFC, 2014a http://www.pefc.org/about-pefc/who-we-are/history (accessed on 29.04.14.).

Penebar Swdaya, T.P., 2008. Panduan Lengkap Karet. Niaga Swadaya.

Pierce, A., Shanley, P., Laird, S., 2003. Certification of non-timber forest products: Limitations and implications of a market-based conservation tool. Paper presented at The International Conference on Rural Livelihoods, Forests and Biodiversity 19–23 May 2003, Bonn, Germany.

Pinchot, 2014. Certification: Definition and Background. http://www.pinchot.org/project/59 (accessed 27.03.14.).

President Decision 43 of 1978 on Convention on International Trade in Endangered Species of Wild Fauna and Flora. Keputusan Presiden No. 43 Tahun 1978.Tentang: Convention On International Trade In. Endangered Species Of Wild Fauna And Flora.

President degree 71 of 1993 on approving Indonesian Agreement on CIFOR's headquarters.Keputusan Presiden No. 71 Tahun 1993 tentang Pengesahan Agreement Indonesia sebagai Kantor Pusat CIFOR.

Presidential Instruction6 of 2013 on New License Suspension and Improving Governance of Primary Forest and Peatlands. Instrusksi Presiden Nomor 6 Tahun 2013 tentang Penundaan Pemberian Izin Baru Dan Penyempurnaan Tata Kelola Hutan Alam Primer dan Lahan Gambut.

Presidential Regulation 62 of 2013 on Agency of the Management of Greenhouse Gas Emission Reduction of Deforestation, Forest Degradation and Peatland.Peraturan Presiden Nomor 62 Tahun 2013 tentang Badan Pengelola Penurunan Emisi Gas Rumah Kaca dari Deforestasi, Degradasi Hutan dan Lahan Gambut.

Prihadi, N., 2013 CITES and its Implementation: case study and experiences of Indonesia. Workshop on Trade & Environment for Asia and Pacific Economies Bogor, 18 September 2013.

Radar Pena, 2014. Moratorium Kehutanan Hambat Investasi Sawit, http://m.radarpena.com/welcome/read/2014/06/04/10906/18/1/Moratorium-Kehutanan-Hambat-Investasi-Sawit (accessed 27.11.14.).

Rainforest Alliance, 2015. An Evaluation of Asia Pulp & Paper's Progress to Meet its Forest Conservation Policy (2013) and Additional Public Statements 18 month Progress Evaluation Report. http://www.rainforest-alliance.org/sites/default/files/uploads/

4/150205-Rainforest-Alliance-APP-Evaluation-Report-en.pdf (accessed 07.04.15)

RAMSARa, 2014. About the Ramsar Convention, http://www.ramsar.org/cda/en/ramsar-about-about-ramsar/main/ramsar/1-36%5E7687_4000_0 (accessed 26.03.14.).

RAMSARb, 2014. Seventh Ramsar Site in Indonesia, http://www.ramsar.org/cda/en/ramsar-activities-wwds-seventh-ramsar-site-indonesia/main/ramsar/1-63-78%5E26389_4000_0 (accessed 26.03.14.).

RAMSAR, 2014c. Ramsar Regional Initiatives, http://www.ramsar.org/activity/ramsar-regional-initiatives

RAN, 2014. http://ran.org/indonesia%E2%80%99s-rainforests-biodiversity-and-endangered-species (accessed 29.04.14.).

Republika, 2014a. Rubber farmers in North Sumatra languishing in poverty. http://www.republika.co.id/berita/en/national-politics/14/03/31/n3agz3-rubber-farmers-in-north-sumatra-languishing-in-poverty (accessed 14.04.14.).

Republika, 2014b. Kemenperin Akan Berikan 100 Sertifikasi SVLK kepada IKM. http://www.republika.co.id/berita/ekonomi/mikro/13/07/02/mpapxz-kemenperin-akan-berikan-100-sertifikasi-svlk-kepada-ikm (accessed 05.04.14.).

Republika, 2014c. Pemerintah Anggarkan Rp 78 Miliar untuk Peremajaan Karet, http://www.republika.co.id/berita/ekonomi/mikro/14/07/17/n8uk3c-pemerintah-anggarkan-rp-78-miliar-untuk-peremajaan-karet (accessed 26.11.14.).

Republika, 2008. Tiga Negara Sikapi Tren Penurunan Harga Karet. http://www.republika.co.id/berita//no-channel/08/11/01/11211-tiga-negara-sikapi-tren-penurunan-harga-karet (Accessed on 26.11.2014.).

REDD-Monitor, 2014. The Warsaw Framework for REDD Plus: The decision on coordination of REDD finance. http://www.redd-monitor.org/2013/12/04/the-warsaw-framework-for-redd-plus-the-decision-on-coordination-of-redd-finance/ (accessed 07.04.2014.).

Rianto, B., Mochtar, H., Sasmito, A., 2012. Overview of Palm oil Industry Landscape in Indonesia.

RSPO, 2014. Why RSPO Certification, http://www.rspo.org/en/why_rspo_certification (accessed 12.03.2014.).

RSPO, 2013. ISPO and RSPO enter into strategic co-operation, http://www.rspo.org/news_details.php?nid=195 (accessed 09.07.14.).

SAN, 2014. http://sanstandards.org/sitio/subsections/display/2 (accessed 03.04.14.).

SAN, 2014. http://sanstandards.org/sitio/subsections/display/1 (accessed 11.03.14.).

Sawit Indonesia, 2014. Diskrimanasi Inpres Moratorium (accessed 27.11.14.).

Sawit Watch, 2007. RSPO Indonesia Liaison Office Telah Dibentuk (21.12.06.).

Siregar C. A., Ridwan M., 2013. Rehabilitasi Lahan di Zona Ekstrim. Belajar dari A/R CDM Lombok Timur. Published by Korea International Cooperation Agency (KOICA), MoFor, CER Indonesia RA Visindo.

Tempo, 2012. Pengusaha Keluhkan Moratorium Lahan Gambut, http://www.tempo.co/read/news/2012/12/20/090449414/Pengusaha-Keluhkan-Moratorium-Lahan-Gambut (accessed 27.11.14.).

Yulisman, L., 2014. Indonesia aims long-term timber export growth. March 01 2014. The Jakarta Post, 2014.

UNCTAD, 2012. International Natural Rubber Agreements (INRA), http://www.unctad.info/en/Infocomm/Agricultural_Products/Caoutchouc/Economic-policies/The-EC-banana-regime/ (accessed on 26.11.14.).

UNFCC, 2014a. LULUCF under the Kyoto Protocol, https://unfccc.int/methods/lulucf/items/4129.php (accessed 26.03.14.).

UNFCC, 2014b. REDD Web Platform, http://unfccc.int/methods/redd/redd_web_platform/items/4531.php (accesed 26.03.14.).

UNORCID, 2014. http://www.unorcid.org/ (accessed 12.03.14.)

UNESCO, 2014a. Global Strategy. http://whc.unesco.org/en/globalstrategy/ (accessed 12.03.14.).

UNESCO, 2014b. Convention Concerning the Protection of the World Cultural and Natural Heritage, http://whc.unesco.org/en/conventiontext/ (accessed on 31.03.14.).

UNESCO, 2002. Universal Declaration on Cultural Diversity A document for the World Summiton Sustainable Development Johannesburg, 26 August–4 September 2002.

WI, 2014a. About Wetlands International, http://www.wetlands.org/Aboutus/tabid/54/Default.aspx (accesed 26.03.14.).

WI, 2014b. http://www.wetlands.org/Whatwedo/Ouractions/tabid/2661/mod/601/articleType/ArticleView/articleId/3558/Default.aspx (accessed 26.03.14).

Wibowo Ari. 2008. Clean Development Mechanism – Forestry and Its Prospect in Indonesia. Tekno Hutan Tanaman. Vol.1, No.1 November 2008, 1–9.

WI-Indonesia, 2014. Peta dan Atlas Distribusi Lahan Gambut, http://indonesia.wetlands.org/Infolahanbasah/PetaSebaranGambut/tabid/2834/language/id-ID/Default.aspx (accessed 02.05.14.).

WRI, 2014. http://www.wri.org/blog/what%E2%80%99s-next-indonesia-norway-cooperation-forests (accessed 12.03.14.).

JMHT Vol. 21, (3): 162–171, December 2015 Scientific Article
EISSN: 2089-2063 ISSN: 2087-0469
DOI: 10.7226/jtfm.21.3.162

Towards State Hegemony Over Agricultural Certification: From Voluntary Private to Mandatory State Regimes on Palm Oil in Indonesia

Muhammad Alif Kaimuddin Sahide[1*], Sarah Burns[1], Agung Wibowo[1], Dodik Ridho Nurrochmat[2], Lukas Giessen[1]

[1]Chair of Forest and Nature Conservation Policy, Georg-August-Universität Goettingen, Buesgenweg 3, Goettingen, Germany, 37077
[2]Department of Forest Management, Faculty of Forestry, Bogor Agricultural University, Academic Ring Road Campus IPB Dramaga, PO Box 168, Bogor, Indonesia 16680

Received October 20, 2015/Accepted November 28, 2015

Abstract

Previous work on certification of palm oil has reported on a trend toward a change, from failed state regulation to voluntary, private governance. However, recent observations suggest a trend, moving from voluntary, private governance to mandatory state governance in palm oil certification in Indonesia, a move in which the state is reclaiming authority. In this light, the aims of our research are (1) to identify the main actors involved in certification politics, (2) to explain this trend in terms of the actors' interests and whatever benefits may result for them. We developed our research questions based on bureaucratic politics and power theory. A mix of document analysis, interviews, and observations are applied for addressing the questions. The results answer our research questions, i.e., that (1) the state claims back its authority over certification from private actors and contributed to the complex meta governance of palm oil certification, the state mandatory scheme that is supported by states' bureaucracies in charge reduces the influence of non-government or private actors. (2) This trend is due to a coalition of specific state bureaucracies and big industry interests, which grant privileges to industry that are denied to small producers. Unexpectedly, all Indonesian bureaucracies associated with this trend support mandatory state certification, which indicates that palm oil has been elevated in importance to become a matter of national, rather than mere bureaucratic interest. Making certification mandatory through coercive regulatory power is the main tool with which state power can challenge voluntary implementation and reclaim authority. Furthermore, the state needs the voluntary system to exist as well in order to strengthen its position. Therefore, the voluntary and the compulsory systems collaborate to attract global initiatives, which is contributing to the high complex of meta governance.

Keywords: international regimes, transnational regimes, RSPO, ISPO, meta governance, palm oil, Indonesia

**Correspondence author, email:alif.mksr@gmail.com, ph.: +49-551-39 13391*

Introduction

Private certification was introduced in the early 1990s to address concerns associated with environmental issues, especially in the tropics (Rametsteiner & Simula, 2003), for many commodities (Blackman & Rivera, 2011). The increased participation by non-state actors through these mechanisms gave way to transnational regimes, differing from state driven international regimes (Pattberg 2012). Transnational regimes include many different issue areas, one example being palm oil certification. The Roundtable on Sustainable Palm Oil (RSPO) is used widely as a reference for private palm oil plantation certification in Indonesia, covering both food and fuel uses (Larsen *et al*. 2014). Palm oil plantations and its industrial sector play an indispensable role for the Indonesian economy bringing more than $21 billion per year in export earnings (Agrofarm 2013b).

Currently, the Indonesian government experiments with state mandatory schemes, through the establishment of the Indonesian Sustainable Palm Oil (ISPO) certification (Edwards & Laurance 2012). A similar transformation from voluntary-private to mandatory-public authority also exists with soybean in Brazil (Hospes 2014) and timber products (Cashore & Stone 2012, Giessen *et al*. 2015, Pratiwi *et al*. 2015, Wibowo *et al*. 2015). The political consequences of the re-establishment of national governance in certification processes have generated several research topics, such as the politics of scaling (Hospes & Kentin 2014), dynamic interactions (Gulbrandsen 2014), intersection (Cashore & Stone 2012) and contested legitimacy (Marin-Burgos *et al*. 2014).

However, there is not enough research in the area to assess state actors' interests in reclaiming their authority,

JMHT Vol. 21, (3): 162–171, December 2015
EISSN: 2089-2063
DOI: 10.7226/jtfm.21.3.162

Scientific Article
ISSN: 2087-0469

especially in the single case of palm oil certification. Therefore, we will explore the interests among the actors of these regulatory (ISPO) and voluntary (RSPO) certification regimes using a bureaucratic (state and non-state) political theory approach that is based on the utility production of these two schemes. Therefore, the aims of this research are (1) to identify the main actors involved in certification politics, and (2) to explain this trend by examining actors' interests and the resulting benefits for them.

Theoretical background

Voluntary private certification Transnational regimes can be defined as "set of norms, rules, and decision-making procedures that are made and implemented across borders through the activities of non-state actors" (Pattberg 2012) differing from international regimes where the states generate the rules (Pattberg 2012). International schemes, as certification, can then be described as transnational regimes which aim to enable the market to control and to select products that comply with quality assurance, offering at the same time greater incentives for industries to implement environmentally and socially responsible objectives (Cashore 2002; McCarthy 2012). This non-state governance has created non-state, market-driven (NSMD), governance systems (Cashore 2002). Cashore *et al.* (2007) argued that "the most important feature of NSMD governance is that there is no use of state sovereignty to enforce compliance". However, Sikor *et al.* (2013) found that new forms of social exclusion, inequity, and ecological simplification have been generated from flow-centred governance, due to the intensive competition for land (including global certification).

McCarthy (2012) analysed the regulation of the palm oil boom in Indonesia, where the RSPO certification system is widely influential, as well as the International Organization for Standardizations (ISO) 14001 series for environmental management systems. Hospes (2014) found that the difficulties in the implementation of this voluntary private standard at the domestic level (in the South) have prompted the collaboration of scholars with national or local governments. The RSPO also provides open space for social and environmental NGOs on their membership board and multi-stakeholder initiatives. Pesqueira & Glasbergen (2013) found that engagement in non-state governance processes gives NGOs valuable advantages, since they contribute to areas to which governments are unwilling or incapable of doing likewise, offering a synergistic and innovative relationship for the creation of benchmarks that can introduce significant changes to current practice, and that encompass sector-wide interventions offering a broader potential for change than collaboration with individual interventions. Busch (2014) argued that this is evidence of both the continuing hegemony of neoliberalism and of various responses to it.

Meta governance: the experiment of mandatory state certification due to bureaucratic politics Currently, a shift towards state certification is observed, suggesting that in pursue of their formal mandates and informal interests state bureaucracies like the Ministry of Agriculture (MoA) and Ministry of Trade (MoT) formally strive for better economic performance by national palm oil industries while at the same time extending their policy domains where they operate, gaining legitimacy, staff and budget. We define the interests of bureaucracies as their search for problem-oriented delivery of public service, and their pursuit of the organisational interests of survival and organisational expansion (Weber 1922; Niskanen 1971; Krott 2005; Peters 2010). A distinctive feature of these bureaucratic politics approach is that bureaucracies in their behaviour follow a dual organizational interest: (1) informally they are competing with other actors and especially other bureaucracies for resources, political domains and influence, and (2) formally, they have distinct tasks for delivering public services. In addition, bureaucray will use their power to reclaim their authority (Giessen *et al.* 2014). In addition, bureaucracies will use their power to reclaim their authority.

However, Following this Weberian approach focusing on bureaucracy and state power has still limitation, this perspective does not grasp the issue of bureaucratic actors operating in their private capacity (e.g. receiving patronage, placing themselves in flows of revenue) (Aspinall & van Klinken 2011), particularly for the very complex situation and land use administration in Indonesia (Sahide & Giessen 2011). Therefore, this actor centred perspective will take into account the informal private interests of bureaucrats, as well as bureaucracies.

We followed this line of argumentation by looking into how the experiment of mandatory public governance reclaims back its authority from the failed global governance of voluntary private certification, which is resulted to the meta-governance due to the high politic of their intersections from both private and state certifications. We define this mandatory public governance as experiment at the moment, as state bureaucracies do not know yet if credibility from NSMD certification also works for state led certification, which only markets will show in the future. Overdevest and Zeitlin (2014) argue that experimentalism provides an analytical framework for evaluating transnational governance interactions in regime complexes where the scope for this experimentalist governance is precisely the opposite of those for regime formation in standard international relations (the regime concept can be seen in section 2.3.).

Mandatory state certification is introduced through legal state verification schemes (Pratiwi *et al.* 2015). This legality can be confirmed by creating formal connections with domestic rules (Schouten & Glasbergen 2011). This scheme is mystifying because it presents a relatively modest solution, compared to previous efforts to build a legally-binding global-private certification system, and yet it is garnering the interest of a wide range of diverse global coalitions within developed and developing countries (Cashore & Stone 2012). The ISPO system is a policy adopted by the Ministry of Agriculture on behalf of the Government of Indonesia and was mandatory from the beginning for all palm oil firms operating in Indonesia (MoA Regulation 19 of 2011). The aim of the ISPO is "to improve the competitiveness of the Indonesian palm oil on the global market and contribute to the objective set by the President of the Republic of

JMHT Vol. 21, (3): 162–171, December 2015
EISSN: 2089-2063
DOI: 10.7226/jtfm.21.3.162

Indonesia to reduce greenhouse gases emissions and draw attention to environmental issues" (ISPO Commission 2013a).

Key research questions Based on the theories above, we develop the following hypotheses around the general assumption that by using certification state bureaucracies claim back authority from private-actors and institutions (Pattberg 2012):

1. Did bureaucracies, for countering private palm oil certification, establish state-driven palm oil certification schemes and useed their regulatory power for even imposing an obligation to certify?
2. How did different bureaucracies favour different certification schemes, e.g. based on bureaucratic interests?. Are the land use bureaucracy in favour of mandatory state schemes whereas the trade bureaucracy favoured private-voluntary schemes?

Methods

To gain our objectives, content analyses was used. Policy documents, the media or professional journals published by political actors are basic resources for qualitative content analysis that is a flexible method for analysing text data (Hsieh & Shannon 2005; Neuman 2005). This literature review is used for identifying actors and their positions in certification politics. In addition, interviews to staff from different ministries and institutions that have broad knowledge of private voluntary and state mandatory palm oil certification, such as the MoA, ISPO Commission, RSPO Indonesian office, and the Palm Oil Enterprise Association (GAPKI) were done. The purpose of these selected interviews was to develop a causal explanation to our hypotheses (Mosley 2013). This paper reviews some of the current main actors and provides descriptions of their interests and power bases. We also delivered observations (during December 2013–February 2014) based on van Evera (1997).

This is also supported by interviews and observations in the field. In the end, we examining actors' interests and the resulting benefits for them of both certification schemes and by explaining who is getting privileges as a result from both state coalition and private coalition, as well as their interests intersection, using bureaucratic politic theory.

Results and Discussion

RSPO-voluntary versus ISPO-mandatory palm oil certification Table 1 shows that private RSPO certification provides tools but cannot address underlying issues that require legal jurisdiction from land use bureaucracies (MoFor, NLA, and Regional Government). This supports McCarty's (2012) argument that certification requires legal reforms.

The RSPO could not introduce their additional criteria for certification, which required companies to go beyond what national law currently requires, because the RSPO does not have a legal orientation (for detailed explanation, it can be seen in Sections The ISPO coalition and RSPO coalition).

The ISPO Coalition

1 Ministry of Agriculture (and ISPO Commission)

The Ministry of Agriculture (MoA) used its coercive power requiring all palm oil companies in Indonesia to certify under ISPO before 31 December 2014 (MoA Regulation 19 of 2011). The MoA applied the ISPO's principles, indicators and standards, which are nearly the same as the ones from RSPO. However, it does not have full coercive power due to the fragmented Indonesian bureaucracies involved with both land use and commodities (Sahide & Giessen 2015). The MoA only has the power to lower a company's classification to fourth grade (the worst). The MoA is increasing its staff and budget, benefitting through a new ad hoc commission under MoA authority, the ISPO Commission, in charge of developing ISPO standards and systems. In one example, the ISPO Commission activated a budget (IDR391,734,124 in 2011) under the Directorate on Postharvest and Business Development of the MoA for ISPO activities such as coordination, consultation, socialisation and system improvement (MoA 2011). At present, the Directorate General for Plantation of MoA is also responsible for palm oil

Table 1 Key features of authority on mandatory-state certification and voluntary-private-certification

	RSPO (Voluntary -private-global certification)	ISPO (Mandatory-public-national verification)
Role of government	RSPO do not require adherence to rules. The MoA recognises RSPO only as a global voluntary scheme.	ISPO formed and governed by MoA (Sovereign governments decide)
Legality orientation	Legality only applied to maintain transparency in global market issues	Legality is the main consideration in a group of rules from many bureaucracies
Authority resources	Market based authority	State based authority
Authorities struggle orientation	Infiltrate domestic-state's land use authority	Infiltrate other countries' trade authority (through international, regional, and bilateral regime)
Degree of Compliance	Voluntary /principle and criteria developed, basedon the consensus of membership or actor's power	Mandatory/legally binding, developed originally by state power.

Source: Adapted from Hospes & Kentin (2014)

JMHT Vol. 21, (3): 162–171, December 2015
EISSN: 2089-2063
DOI: 10.7226/jtfm.21.3.162

Scientific Article
ISSN: 2087-0469

development, which traditionally has had a large budget and staff. Now, the MoA can develop any changes, or adaptations, and can also use a 'multi-stakeholder' approach based on global trends and markets. Furthermore, the MoA has asked the RSPO to accept the ISPO as the complement of the RSPO standard (Antara 2013). To support strong legality issues on palm oil area, MoA has issued MoA Regulation of 2013 on the Guideline of Plantation Businesss. This rules limiting the maximum area of 100 thousand hectares of every company.

2 National Land Agency (NLA)
As one of the core bureaucracies that play indispensable roles in providing land for palm oil plantation (Sahide & Giessen 2015) as well as NLA rules, it has been set up as one of the major contribution to the legality indicators of the ISPO standard. The NLA will use any legal position to strengthen their domain responsibilities, especially around providing land for plantations, and this will affect the increased budget for any activities related to this issue, particularly activities in the field. For example, the NLA is not regarded as being fully supportive of the implementation of HCV from the RSPO standard. Employees of a big palm oil company have confirmed that the NLA sent a warning letter to the private palm oil company after the company applied HCV in their area. The NLA argued that all land concession should be managed, otherwise it might be categorised as wasteland, which is against the law. The MoFor support this argument that if the company consider HCV is a wasteland and then to be open access resources, the problem is how government can guarantee HCV area will work (it assume when HCV is needed for biodiversity protection) in absent of manager of the land).

3 Ministry of Forestry (MoF)
The MoF rules also contribute to the legality indicator for the ISPO and it plays a substantial role in releasing forest area for palm oil plantations (Sahide & Giessen 2015). The MoF will benefit from ISPO certification as a legality tool, using it as a reason for increasing their budget to campaign against illegal palm oil plantation activities in forest areas or to campaign so that ISPO certification cannot be considered for planting palm oil in state forest areas based on the legality principles, the MoF will only fully support ISPO implementation challenging RSPO standard implementation, which is inline with Niskanen (1971) that one of the bureaucracy interests is maximizing budget. For example, applying the HCV standard of the RSPO to a given palm oil concession area would lead to it being categorised as wasteland, which is against the Government Regulation 11 of 2010 on Controlling and Using Waste Land. Consequently, the palm oil concession right would be revoked by the National Land Agency (Agrofarm 2014). The RSPO's prohibition against the development of any new plantation after November 2005 by replacing natural forest or areas required to maintain or enhance high conservation value is known controversially within the plantation industry as the "cut-off-date" and goes against the MoF who 'regularly' release state forest area for plantation purposes (MoF 2011; 2012; 2013).

4 Regional Government
The regional government includes important bureaucracies that issue preliminary licences to provide land for palm oil concession (Sahide & Giessen 2015) They will benefit from the fact that ISPO certification will increase their political responsibilities, and they expect a regional income from palm oil plantations. The regional government comprises the central state actors at the local level that establish the legality element of the ISPO standard.

5 The Ministry of Environment (MoE)
Regulates the spatial planning and Environmental Impact Assessment processes that aim to ensure that the environmental principles are applied to the planning process and that identify and address the anticipated environmental effects of palm oil plantation activities. ISPO provides a complete standard for the production and long term viability of the plantation organisation, but regarding conservation and protection of rare, threatened and endangered species, the ISPO relies on the national environmental assessment, which is governed by the MoE, and become a major legality indicator of the ISPO standard. However, the credibility and capability of assessors on the Environmental Impact Assessment platform are sometimes criticised by non-state actors.

6 Big palm oil producer companies
Big palm oil companies that operate in Indonesia, like PT Bakrie Sumatra Plantation (BSP) and PT London Sumatra Plantation, play important roles because of their dual membership, being registered in both the RSPO membership and the GAPKI, which supports ISPO (RSPO 2012; GAPKI 2013). The basic motive for this is that those companies are in a strategic position to acquire information among the RSPO and ISPO coalitions and to use it to benefit from both schemes.

7 The Ministry of Trade (MoT)
Supports fiscal policies, such as planning to ban exports of palm oil without ISPO certificates (Medan Bisnis Daily 2013). They support the MoA establishing and requiring mandatory ISPO certification (Sindo News 2014). The MoT will also provide incentives for ISPO certificate holders, like export tax reductions (BUMN 2014). The MoT and MoA are lobbying in Europe and other countries for the acknowledgement of ISPO certification. The MoT is negotiating ISPO in global trade forums like the World Trade Organisation since they are in charge, as the focal point of the World Trade Organisation (MoFA 2014). The Vice Minister of Trade argued that Indonesian palm oil companies that have the ISPO certification do not need the recognition of the RSPO certification (Nasionalisme 2014).

8 Ministry of Industry (MoI)
The development of industries downstream of the palm oil industry in Indonesia must be improved in order to increase added value. Indonesia is currently only able to produce about 47 derivative products from palm oil (MoI

2012). This situation makes the MoI's position an important one in coalition with the MoA.

9 GAPKI (Indonesian Palm Oil Enterprise Association)
This Indonesian private-business organisation was formed in 1981 (GAPKI 2014). As of 2013, GAPKI has 591 members, with a corresponding palm oil plantation area of around 3 million hectares, more than 30% of the Indonesian palm oil area, being an important government partner in improving the situation of Indonesian palm oil. GAPKI is also imposing the implementation of the Indonesian Sustainable Palm Oil (ISPO), which empowers Indonesia's sustainability-related regulation of palm oil production, especially for its members. In the beginning, GAPKI was registered as a member of the RSPO, but some GAPKI members argued that even though producers make up the biggest proportion of RSPO's membership, producer interests cannot be accommodated in the RSPO. Therefore GAPKI announced its resignation from the RSPO in 2011, and supported ISPO. The MoA used this momentum to strengthen the ISPO and may have complicated the national implementation of RSPO principles and criteria in Indonesia. GAPKI is an important actor that possesses power and influence to shape the ISPO as an instrument to reclaim authority over palm oil certification. GAPKI, as a non-state actor, has a limited formal ability to drive MoA outcomes, but it has ample capacity to enhance 'ISPO power'.

10 APKASINDO (Association of Indonesian palm oil farmers)
APKASINDO is a non-profit organization fighting for the rights of small-scale plantations in Indonesia (Exposenews 2009). In 1999, APKASINDO jointly with GAPKI sent a protest letter to RSPO after RSPO included reduce greenhouse gases as their indicator of assessment (Exposenews 2009). Even though this organisation supports ISPO they criticize the restrictions policy of concession palm oil business which only managed a maximum of 100,000 hectares (Ministry of Agriculture Regulation of 2013), claiming it will reduce investment of the palm oil sector affecting negatively the farmers (Sawit Indonesia 2013). This means also that this organisation mostly present their motives on taking advantages from governments actors and also private business actors. APKASINDO would like to strengthen GAPKI position on reinforcing ISPO's coalition for taking benefits from the beginning and rich ISPO incentives from government.

11 RSPO
RSPO respect domestic rules and regulations, including ISPO systems, therefore facilitating its Indonesian members' compliance with national rules, including ISPO systems. The Indonesian office of the RSPO will also facilitate application to the ISPO for its Indonesian firm members, they support ISPO for structural reasons not based on the basic interest.

11 The UNDP (United Nations Development Programme) is successfully connecting the RSPO and the ISPO through a joint study programme (ISPO Commission 2013b; RSPO 2013). They also endorse the Sustainable Palm Oil initiative, along with Indonesian actors. This program will be coordinated by key actors such as the Ministry of Forestry, the Ministry of Environment, the Ministry of National Planning, Presidential Unit for Supervision and Development, the Agency of REDD+ (reduce emissions from deforestation and forest degradation, and foster conservation, sustainable management of forests, and enhancement of forest carbon stocks), the Ministry of Cooperatives and Small and Medium Enterprises, the National Land Agency, the Ministry of Trade, the National Standardization Agency, provincial and district governments, the private sector, smallholders, relevant palm oil associations, civil society (ISPO Commission 2013c). This initiative will be included or withdrawn from ISPO criteria after consideration by the ISPO Commission.

12 Indonesian Palm Oil Board (DMSI)
This institution is a non-state actor established based on article 19 Law Number 18 Year 2004 on Plantation. The goal of this institution is to mediate actors' interests to support palm oil for national income (DMSI 2009). Six palm oil enterprise associations and two directorate generals of the MoA are listed among these institutions. This institution focus on the industrialisation of palm oil, and the MoA supports them. The secretariat of this organisation is also housed in a MoA building near the ISPO Commission office. Information elements mean that this institution has power to conduct their formal responsibilities.

13 Indonesian Association of Biofuel Producers
This is a non-state actor, established to support the mandatory policy of the 10% biodiesel (Okezone 2013). Therefore this institution is allied to both the MoEMR and the MoA. Its element of power is the limited information on the enterprise of biofuels.

14 Auditor
They will benefit from the increased demand for both certification schemes. Some auditor providers prefer ISPO over RSPO because their indicator is much easier to be interpreted due to its indicator mainly to the domestic laws, so the Auditor will have a flexibility to decide and do not have much pressure from the state. To get certificate as an auditor, individual or company should pay about IDR3,500,000 until IDR16,000,000, which is conducted by ISPO Commission or other legalized agencies (ISPO Commission 2015; Training Centre 2015). Fundamentally, auditors work independently, but they have the discretion to test the 'doubtful' indicators (like social and environmental indicators) used in the assessment process, so there is a certain likelihood that they will provide a subjective assessment and even that their results could be affected by other actors' motives (Silva-Castañeda 2012). Only 7 independent audit organisations registered in 2003 and approximately 11 in 2014 accessed around 1200 palm oil firms in Indonesia (Agrofarm 2013a; East Kalimantan Plantation Agency 2014). Both RSPO and ISPO require an independent

auditor. Remarkably, some of the independent auditor organisations (like PT Sucofindo Persero, PT Mutuagung Lestari, PT TUV Rheinland Indonesia and PT SAI Global Indonesia) assess both RSPO and ISPO proposals for palm oil firms. Each has approximately 10 auditors.

The RSPO Coalition

1 RSPO
This global, multi-stakeholder organization's membership is comprised of 50% palm oil growers, 26% from palm oil processors and traders, 17% from organisations, 4% from conservation and environmental NGOs, 3% from social and developmental NGOs, and 2% from good manufacturers (RSPO Indonesia Office 2014). This shows that palm oil growers and producers make up the biggest portion of the RSPO's membership, which is contrary to the Indonesian actors' assumption, i.e., that RSPO is much more representative of consumers′ interests.

2 Big palm oil consumers
Indonesian bureaucracies perceive the RSPO to be shaped mainly by consumer interests (Agrofarm 2014). Big palm oil consumers like Unilever, Anone, and Vollmar GmbH have membership in the RSPO (RSPO 2014a). Unilever announced their commitment to sustainable palm oil in the mid-1990s. It would seem that big palm oil consumer companies have settled their business and have chosen the voluntary scheme, due to the involvement and backing of environmental and social NGOs. The involvement of these NGOs contributed to the market legitimacy to strengthen demand actors' (big palm oil consumers) motives and positions to better struggle with supply actors′ (Indonesian big palm oil producer) interests. Interestingly, this big palm oil company is not bothered with palm oil occupancy policy with a maximum of 100,000 ha and some other administrative and legality requirements. For example, SIPEF group who has a majority stake in 14 palm oil companies which is operated around 90 years in Indonesia through their Managing Director argued that this policy is not a big problem because 100,000 ha of palm oil is not easy any more to find in the recent plantation situation in Indonesia (Sawit Indonesia 2013c).

3 The commission of the European Union
Most European countries see climate change as a very serious problem. The European public has recognised the RSPO scheme as a respectable form of voluntary certification, particularly since RSPO RED would be the first voluntary certification to address climate change issues with regard to palm oil or this 'RSPO-RED has been approved by European Commission as a certification scheme accepted in compliance with the requirements of EU RED (Renewal Energy Directive) (RSPO 2014c). However, EU countries prefer RSPO over ISPO to ensure palm oil from Indonesia is 'efficiently' entering European countries, or a good strategy to challenge large numbers and cheap palm oil from Indonesia.

4 Environmental and social NGOs
We identified several NGOs that have strong concerns about serious unintended social and indigenous issues, and economic and environmental implications such as WWF, Rainforest Alliance, Oxfam and Sawit Watch (RSPO 2014b). Greenpeace Indonesia is an environmentalist NGO that exposes environmental problems and uses direct action, lobbying and research to achieve its goals. Greenpeace focuses its work on worldwide issues such as global warming and deforestation (Greenpeace 2011). Sawit Watch is a national NGO that chose to join the RSPO. Sawit Watch has been critical about environmental, human rights violations, labour and social issues (Sawit Watch 2014). They are very vocal and their message can often be distorted so as to be seen to be anti-palm, which confuses the issues sometimes.

5 Ministry of Agriculture (MoA)
The MoA also formally supports voluntary RSPO schemes for certain market purposes (Agroconsultant 2014). The Vice Minister of Agriculture said that the government does not deny the existence of the RSPO certification and that it is supposed to collaborate with the formation of a single palm oil certification in Indonesia (Bewara 2014). The Director of RSPO Indonesia Office proposed RSPO-ISPO cooperation starting with a joint audit. In addition, the Director General of Plantation of the MoA said the proposed joint audit of the certification process of the RSPO with that of the ISPO has the potential to reduce costs significantly (Tempo 2011). The former Minister of Forestry and Agriculture disagrees with some Indonesian actors that do not support the RSPO. He has now become an advisor on the RSPO board (RSPO 2011). This also shows that the RSPO tries to place an Indonesian ex-high level officer in a position that would strengthen their influence. In the informal mission, MoA hope that ISPO can be a pre requirement for RSPO standard (Antaranews 2013).

6 UNDP as well as Auditor
is also observed supporting RSPO coalition, and the detailed of this (actor's explanation has been described on the previous section (see Section ISPO coalition)

State certification established by regulatory power A bundle of regulations from the Ministry of Agriculture, the Ministry of Forestry, the National Land Agency, the Ministry of Environment was set up as a legal tool for ISPO standards and principles. This legality mechanism (Schouten & Glasbergen 2011) is successful in. collecting all power elements (coercive, incentive, and informational) from fragmented Indonesian bureaucracies to support the ISPO (Weber 1922; Krott *et al.* 2014). The use of regulatory power is similar with SVLK in Indonesian forestry sector (Wibowo *et al.* 2015 and Pratiwi *et al.* 2015). Even though Indonesia uses intensive regulatory power, it doesn't mean automatically have higher global legitimacy, but increase legitimacy contestation with RSPO (Marin Burgos *et al.* 2014).

The meta governance system: Bureaucratic interests explaining who is favouring voluntary and mandatory and its intersection We expected through our research question that MoA favour ISPO schemes while MoT favour RSPO since the MoA created ISPO and MoT needs RSPO's existence for their successful negotiation of international certified palm oil. Surprisingly, all Indonesian bureaucracies support ISPO. This phenomenon shows that palm oil and ISPO have become a 'national interest' in Indonesian political bureaucracy. The ISPO, from the perspective of Indonesian bureaucracies, appears to be a tool serving the interests of producers, whereas the RSPO would seem to serve the consumers.

In addition, a new phenomenon shows that RSPO and ISPO are entering a collaboration effort that is also supported by International actors like the UNDP. Joint studies audits and even proposals for the formation of single certification schemes are indicative of both schemes assessing the situation and making decisions when this benefits both schemes, similar with Gulbrandsen (2014) that state-private relation has increase dynamic interactions but different on the politic of scaling (Hospes & Kentin 2014).

Indonesian ministries seem to fully support the ISPO while they seem not to be against the RSPO. However, we also found that there is no 'real' actor opposing both the ISPO and the RSPO. These results show that pro ISPO and pro RSPO actors formally 'support' each other to shape their informal interests, that is contributed to the complex meta governance system. This phenomenon would lead the RSPO to assert their legitimacy over that of the ISPO in the future, which is in line with the argumentation by Overdevest & Zeitlin (2012), that private certification will take back its legitimacy due to the trend from failed public governance to private experimentalism. In the end, the voluntary and the compulsory systems collaborate to attract global initiatives, shaping state and private interests. The ambiguity relations between states also indicate this meta-governance and private actors on making this jointly efforts, Mandatory certification have a critical orientation problem. If government orientation for protecting communities from environmental disaster, it can be accepted. But, if orientation of certification for legality of everything, its rather ambiguous, because if any certain company operate in certain country and the operation is illegal, it shows that the government cannot control their authority.

Non-state actors seem not to have enough roles in the state mandatory ISPO scheme. However, the RSPO accommodates them (like Sawit Watch and Greenpeace Indonesia). The involvement and backing up from NGOs as 'third pressure' contributed to the market legitimacy to strengthen RSPO' alliance. However, the ISPO is very confident since all state bureaucracies as well as some non-state actors, like business and farmer associations support ISPO. This 'global-market-legitimisation' has a new a big challenge from 'domestic-state-legitimation' when the ISPO is introduced to attract powerful 'foreign states' and forces them to collaborate. This result supports our hypothesis that state mandatory schemes supported by state bureaucracies limit the influence of NGOs or private actors.

Conclusion

'Reclaiming' is the bureaucratic power process, shaped by state actors' interests, strongly and more successfully claiming back certification authority from private actors and show the trend of state hegemony. It is inline with Hospes & Kentin (2014) that there is a significant role of state in palm oil certification. The state seems to force voluntary schemes to accept ISPO as complementary to the RSPO standard. This is ambiguous because some ISPO and RSPO standards conflict in the implementation of their principles especially on legal issues. This not only enforces the reclaiming process but also successfully establishes powerful standards through legal means that affirm the global trade governance mechanism. As the ISPO is not really replaces the RSPO, or ISPO intersect with RSPO has resulted to the complex palm oil meta governance similar with (Cashore & Stone 2012) on investigating this private-state intersection. Finally, ISPO has to jointly collaborate with RSPO under UNDP facilitation to enlarge global market cover and increase their credibility over black campaign on illegal palm oil resources as well as social and environmental issues, it is inline with Willetts (2001) that governments are losing dominance when faced with the economic activities of transnational companies and the violent issues (Willetts 2001). In addition, we can observe how the complex meta governance becomes less complex, as national policies interconnect international institutional elements of the palm oil certification regime complex. We may also perceive it as 'less fragmentation" which is derived by the formal and informal cooperation from both coalition in the practical. The sovereignty of Indonesian actors to formulate mandatory certification has reduced the external interest pressure and provides flexibility to define each standard and indicator. Rather than fragmentation in private certification schemes, competition between private and public mandatory schemes resulted in their mutual adjustment and the increased accountability of schemes, which were simplified under the forms of legality of domestic sovereignty. Hence, our hypothesis that setting standards and indicators that serve the informal interest of state bureaucracies contribute to a less fragmented Indonesian tropical rainforest transformation regime complex is proved. Governing the government is the character of reclaiming tools and state hegemony, which will have a good impact on the global credibility if they can follow it thoroughly in the principle of good governance. This can be assessed in the future. Some strategic policy can be recommended such as strengthening the transparency of Environmental Impact Assessment and increasing public participation on monitoring ISPO implementation both in the policy making process and at application in the field. Furthermore, this 'national interest' is potentially to become a regional interest through ASEAN (Association of Southeast Asian Nations) regime, which is led by Indonesia, Malaysia, and Thailand which is now preparing for ASEAN sustainable palm oil certification schemes, and this domestic certification could potentially became a tools for restructuring palm oil industry sector to face ASEAN economic community.

Acknowledgements

We greatly appreciate to financial support from DFG research project on the fragmentation of the international forest regime complex (PAK 813), the Eva Mayr-Stihl Foundation as well as Indonesian Government fund (DIKTI scholarship).

References

Agroconsultant. 2014. *Peran Strategis ISPO dalam bisnis produk kelapa sawit*.

Agrofarm. 2014. *Perusahaan Kurang Siap, Susah Memperoleh ISPO*. http://agroconsultant.asia/perastrategis-ispo-dalam-bisnis-produk-kelapa-sawit.html. [04 June 2014].

Agrofarm. 2013a. *ISPO Kekurangan Auditor*. http://www.agrofarm.co.id/read/sawit/241/ispo-kekurangan-auditor/. [04 June 2014].

Agrofarm. 2013b. *Indonesia Sungguh-Sungguh Menerapkan Standar Sustainability Palm Oil*. http://www.agrofarm.co.id/read/sawit/264/indonesia-sungguhsungguh-menerapkan-standar-sustainability-palm-oil/#.U5AwGPmSxqU. [04 June 2014].

Antaranews. 2013. *Kementan berharap ISPO jadi syarat RSPO*. http://www.antaranews.com/berita/374033/kementan-berharap-ispo-jadi-syarat-rspo. [16 July 2014]

Aspinall E, Van Klinken G. 2011. *The State and Illegality in Indo*. Netherland: Brill. http://dx.doi.org/10.1163/9789004253681_002.

Bewara. 2014. *SERTIFIKASI CPO, Kementan: Ada Peluang Digabungkan*. http://bewara.co/read/2014/05/sertifikasi-cpo-kementan-ada-peluang-digabungkan/ , Accessed on [17 July 2014]

Blackman A, Rivera J. 2011. Producer‐Level Benefits of Sustainability Certification. *Conservation Biology* 25(6):1176–1185. http://dx.doi.org/10.1111/j.1523-1739.2011.01774.x.

BUMN. 2014. Perusahaan Sawit akan Mendapat Insentif. http://www.bumn.go.id/ptpn5/berita/10664/Perusahaan.sawit.akan.dapat.insentif. [04 June 2014].

Busch L. 2014. Governance in the age of global markets: challenges, limits, and consequences. *Agriculture and Human Values* 31(3):513–527. http://dx.doi.org/10.1007/s10460-014-9510-x.

DMSI. 2009. http://dmsi.or.id/aboutus. [16 July2014].

Cashore B. 2002. Legitimacy and the privatization of environmental governance: How non–state market–driven (NSMD) governance systems gain rule–making authority. *Governance* 15(4):503–529. http://dx.doi.org/10.1111/1468-0491.00199.

Cashore B, Egan E, Auld G, Newsom D. 2007. Revising theories of nonstate market-driven (NSMD) governance: lessons from the Finnish forest certification experience. *Global Environmental Politics* 7(1):1–44. http://dx.doi.org/10.1162/glep.2007.7.1.1.

Cashore B, Stone MW. 2012. Can legality verification rescue global forest governance? Analyzing the potential of public and private policy intersection to ameliorate forest challenges in Southeast Asia. *Forest policy and economics* 18:13–22. http://dx.doi.org/10.1016/j.forpol.2011.12.005.

Edwards DP, Laurance SG. 2012. Green labelling, sustainability and the expansion of tropical agriculture: critical issues for certification schemes. *Biological Conservation* 151(1):60–64. http://dx.doi.org/10.1016/j.biocon.2012.01.017.

East Kalimantan Plantation Agency. 2014. http://disbun.kaltimprov.go.id/berita-497-penerapan-ispo-mundur-desember-2015.html. [25 June 2014].

Exposenews. 2009. *Gapki dan Apkasindo Protes RSPO*. http://eksposnews.com/view/7/8477/Gapki-dan-Apkasindo-Protes-RSPO.html. [04 August 2014].

GAPKI. 2013. http://www.gapki.or.id/Page/Gapki Members?page=4.[30 June 2014].

GAPKI. 2014. http://www.gapki.or.id/Page/About. [04 June 2014].

Giessen L, Krott M, Möllmann T. 2014. Increasing representation of states by utilitarian as compared to environmental bureaucracies in international forest and forest–environmental policy negotiations. *Forest Policy and Economics* 38:97–104. http://dx.doi.org/10.1016/j.forpol.2013.08.008.

Giessen L, Burns S, Sahide MAK, Wibowo A. 2015.. From transnational governance to international government regimes in forest and agricultural certification: The key role of state bureaucracies for private governance in Argentina and Indonesia. Policy & Society, under revision.

Gulbrandsen LH. 2014. Dynamic governance interactions: Evolutionary effects of state responses to non‐state certification programs. *Regulation & Governance* 8(1):74–92. http://dx.doi.org/10.1111/rego.12005.

Hospes O, Kentin A. 2014. Tensions Between Global‐Scale and National-Scale Governance: The Strategic Use of Scale Frames to Promote Sustainable Palm Oil Production in Indonesia. *Scale-Sensitive Governance of the Environment* 203.

Hospes O. 2014. Marking the success or end of global multi-stakeholder governance? The rise of national sustainability standards in Indonesia and Brazil for palm

JMHT Vol. 21, (3): 162–171, December 2015 Scientific Article
EISSN: 2089-2063 ISSN: 2087-0469
DOI: 10.7226/jtfm.21.3.162

oil and soy. *Agriculture and Human Values* 1-13. http://dx.doi.org/10.1007/s10460-014-9511-9.

Hsieh HF, Shannon SE. 2005. Three approaches to qualitative content analysis. *Qualitative health research* 15(9):1277-1288. http://dx.doi.org/10.1177/1049732305276687.

ISPO Commission. 2013a. http://www.ispo-org.or.id/index.php?option=com_content&view=article&id=93&Itemid=271&lang=en. [25 June 2014].

ISPO Commission. 2013b. http://www.ispo-org.or.id/index.php?option=com_content&view=article&id=93&Itemid=271&lang=en. [25 June2014].

ISPO Commission. 2013c. http://www.ispo-org.or.id/index.php?option=com_content&view=category&layout=blog&id=14&Itemid=219&lang=en, Accessed on 25.06.2014.

ISPO commission, 2015. http://www.ispo-org.or.id/index.php?option=com_content&view=article&id=111:auditor-ispo-angk-16&catid =13&Itemid=213&lang=ina. [21 November 2015].

Krott M, Bader A, Schusser C, Devkota R, Maryudi A, Giessen L, Aurenhammer H. 2014. Actor-centred power: The driving force in decentralised community based forest governance. *Forest Policy and Economics* 49:34–42.

Krott M. 2005. *Forest Policy Analysis*. Dordrecht: Springer.

Larsen RK, Jiwan N, Rompas A, Jenito J, Osbeck M, Tarigan A. 2014. Towards 'hybrid accountability'in EU biofuels policy? Community grievances and competing water claims in the Central Kalimantan oil palm sector. *Geoforum* 54:295–305. http://dx.doi.org/10.1016/j.geoforum.2013.09.010.

Marin-Burgos V, Clancy JS, Lovett J C. 2014. Contesting legitimacy of voluntary sustainability certification schemes: Valuation languages and power asymmetries in the Roundtable on Sustainable Palm Oil in Colombia. Ecological Economics.

Medan Bisnis Daily. 2013. Ekspor Sawit Tanpa ISPO akan Dilarang. http://medanbisnisdaily.com/news/read/2013/11/13/61507/ekspor_sawit_tanpa_ispo_akan_dilarang/ , [01 July2014].

McCarthy JF. 2012. Certifying in Contested Spaces: private regulation in Indonesian forestry and palm oil. Third World Quarterly 33(10):1871–1888. http://dx.doi.org/10.1080/01436597.2012.729721.

[MoA] Ministry of Agriculture. 2011. http://deptan.go.id/sakip/admin/data2/LAKIP_DIR_PASCAPANEN_2011.pdf. [09 July 2014].

MoA Regulation 19 of 2011 on ISPO as mandatory for palm oil plantation firm in Indonesian. *Peraturan Menteri Pertanian No. 19/Permentan/OT.140/3/2011 tentang Kewajiban Sertifikasi ISPO Bagi Perusahaan Perkebunan*. Jakarta: Ministry of Agriculture.

MoI. 2012. http://kemenperin.go.id/artikel/6372/Industri-Sawit-Harus-Dibedakan-untuk-Makanan-dan-Nonmakanan. [16 July 2014].

MoFA. 2014. *Keanggotaan Indonesia pada Organisasi Internasional*. http://kemlu.go.id/Documents/Keanggotaan_Indonesia_pada_OI.pdf. [26 March2014].

[MoF] Ministry of Forestry. 2011. Indonesian Forestry Statistic in 2010. Jakarta: Ministry of Forestry.

[MoF] Ministry of Forestry. 2012. Indonesian Forestry Statistic in 2011. Jakarta: Ministry of Forestry.

[MoF] Ministry of Forestry. 2013. Indonesian Forestry Statistic in 2012. Jakarta: Ministry of Forestry.

Mosley L. 2013. Introduction. "Just Talk to People"? Interviews in Contemporary Political Science. In: Mosley L, Editor. *Interview Research in Political Science*. London: Cornell University Press.

Nasionalisme. 2014. http://www.nasionalisme.co/?p=895. [18 July 2014].

Neuman WL. 2005. Social research methods. Quantitative and qualitative approaches. 6th edition. London, UK Allyn and Bacon.

Niskanen W. 1971. Bureaucracy and Representative Government. Aldine-Atherton, Chicago

Okezone. 2013. Asosiasi Biofuel Dukung Kebijakan 10% Biodiesel. http://economy.okezone.com/read/2013/09/17/19/867622/asosiasi-biofuel-dukung-kebijakan-10-biodiesel, Accessed on 16.07.2014.

Overdevest C, Zeitlin J. 2012. Assembling an experimentalist regime: Transnational governance interactions in the forest sector. *Regulation & Governance* 8(1):22–48.

Pattberg P. 2012. Transnational Environmental Regimes. In: F. Biermann, P. Pattberg, editors. *Global Environmental Governance Reconsidered*. Cambridge: MIT Press.

Pratiwi S, Wibowo A, Giessen L. 2015. Third-Party Certification of Forest Management In Indonesia: Analysing Stakeholders' Recognition and Preferences. *Jurnal Manajemen Hutan Tropika* 21(2):65–75. http://dx.doi.org/10.7226/jtfm.21.2.65.

Pesqueira L, Glasbergen P. 2013. Playing the politics of scale: Oxfam's intervention in the Roundtable on Sustainable Palm Oil. *Geoforum* 45:296–304. http://dx.doi.org/10.1016/j.geoforum.2012.11.017.

Peters BG. 2010. The Politics of Bureaucracy-An Introduction to Comparative Public Administration. Sixth edition. Routledge. Oxon

JMHT Vol. 21, (3): 162–171, December 2015
EISSN: 2089-2063
DOI: 10.7226/jtfm.21.3.162

Scientific Article
ISSN: 2087-0469

Pratiwi S, Wibowo A, Giessen L. 2015. Third-party certification of forest management in Indonesia: Analysing stakeholders' recognition and preferences. *Journal Manajemen Hutan Tropika* 21(2):65–75. http://dx.doi.org/10.7226/jtfm.21.2.65.

Rametsteiner E, Simula M. 2003. Forest certification–an instrument to promote sustainable forest management? *Journal of environmental management* 67(1):87–98. http://dx.doi.org/10.1016/S0301-4797(02)00191-3.

Raustiala K, Victor DG. 2004. "The Regime Complex for Plant Genetic Resources," International Organization 55:277–309. http://dx.doi.org/10.1017/s00208183045 82036.

RSPO. 2014a. http://www.rspo.org/en/certified_grower, [04 June 2014].

RSPO. 2014b. http://www.rspo.org/en/member/listing/category/Consumer%20Goods%20Manufacturers. [16.July 2014].

RSPO. 2014c. http://www.rspo.org/en/rspo-red_briefing_for_members. [14 August 2014].

RSPO. 2013. ISPO and RSPO enter into strategic co-operation. http://www.rspo.org/news_details.php?nid=195. [09 July 2014].

RSPO. 2012. http://www.rspo.org/en/member/23/pt-pp-london-sumatra-indonesia-tbk. [30 June 2014].

RSPO. 2011. *RSPO Tunjuk Mantan Menteri Pertanian dan Kehutanan Bungaran Saragih Sebagai Penasehat.* http://www.rspo.org/news_details.php?nid=13&lang=5. [17 July 2014].

Sahide MAK, Giessen L. 2015. The fragmented land use administration in Indonesia–Analysing bureaucratic responsibilities influencing tropical rainforest transformation systems. *Land Use Policy* 43:96–110. http://dx.doi.org/10.1016/j.landusepol.2014.11.005.

Sawit Indonesia. 2013. *Investasi Hulu Sawit Semakin Tertekan* 2(25): 18–20

Sawit Watch. 2014. http://sawitwatch.or.id/2011/09/sawit-watch-and-rspo/#sthash.OmrmjTFB.dpuf. [04 July 2014.

Sindo News. 2014. http://ekbis.sindonews.com/read/856224/34/kemendag-gencarkan-sertifikasi-ispo-untuk-petani-sawit. [09 July 2014].

Schouten G, Glasbergen P. 2011. Creating legitimacy in global private governance: The case of the Roundtable on Sustainable Palm Oil. *Ecological economics* 70(11):1891–1899. http://dx.doi.org/10.1016/j.ecolecon.2011.03.012.

Sikor T, Auld G, Bebbington AJ, Benjaminsen TA, Gentry BS, Hunsberger C, Izac A, Margulis ME, Plieninger T, Schroeder H, Upton C. 2013. Global land governance: from territory to flow? *Current Opinion in Environmental Sustainability* 5(5):522–527. http://dx.doi.org/10.1016/j.cosust.2013.06.006.

Silva-Castañeda L. 2012. A forest of evidence: third-party certification and multiple forms of proof—a case study of oil palm plantations in Indonesia. *Agriculture and Human Values* 29(3):361–370. http://dx.doi.org/10.1007/s10460-012-9358-x.

Tempo. 2011. *Pemerintah Pertimbangkan Audit Gabungan Sertifikasi Sawit.* http://bisnis.tempo.co/read/news/2011/05/19/090335482/pemerintah-pertimbangkan-audit-gabungan-sertifikasi-sawit [17 July 2014].

Training Centre. 2015. http://www.trainingcenter.co.id/audit-rspo-ispo. [21November 2015].

Van Evera S. 1997. Guide to methods for students of political science. Cornell University Press.

Weber M. 1922. Wirtschaft und Gesellschaft. Grundriß der verstehenden Soziologie. Studienausgabe. Zitiert nach der Ausgabe von 1988. Tübingen.

Wibowo A, Sahide MAK, Dharmawan B, Pratiwi S, Giessen L. 2015. Global forest certification schemes in Indonesia and their transformation options. *Risalah Kebijakan Pertanian dan Lingkungan* 2(1): 1–8.

Willetts P. 2001. Transnational actors and international organizations in global politics. The globalization of world politics, 2.

Ref.: FORPOL2213
Title: Is Indonesia utilising its international partners? The driving forces behind Forest Management Units
Authors: Muhammad Alif Kaimuddin Sahide; Ahmad Maryudi, Dr; Supratman Supratman, Prof. Dr.; Lukas Giessen, Dr

Dear Mr. Sahide,

I can now inform you that the Editorial Board has evaluated the manuscript FORPOL2213.

The Editor has advised that the manuscript will be reconsidered for publication after major revision.

In order to facilitate a quick editorial process you are kindly requested to submit a revised version of your manuscript within 2 months.

Please visit your ees account in order to view the
comments. Along with your revised manuscript, you will need to supply a covering letter in which you list all the changes you have made to the manuscript, and in which you detail your responses to all the comments passed by the reviewer(s) and the Editor. Should you disagree with any comment(s), please explain why.

To submit a revision, please visit http://ees.elsevier.com/forpol/ and log in with the user name and password supplied earlier, selecting the 'Author Login' button. You will see a menu item called Submissions Needing Revision. The revised manuscript and covering letter can be submitted there.

When submitting your revised manuscript, please ensure that you upload the source files (e.g. Word). Uploading a PDF file at this stage will create delays should your manuscript be finally accepted for publication. If your revised submission does not include the source files, we will contact you to request them.

To speed up the production process, I would like to urge you to upload all source files separately. Figures should be uploaded per number. For example, Figures 1, 2a, 3 should be uploaded as 3 separate files.

Please proceed to the following link to update your personal classifications and keywords, if necessary:

Please note that this journal offers a new, free service called AudioSlides: brief, webcast-style presentations that are shown next to published articles on ScienceDirect (see also http://www.elsevier.com/audioslides). If your paper is accepted for publication, you will automatically receive an invitation to create an AudioSlides presentation.

Forest Policy and Economics features the Interactive Plot Viewer,
see: http://www.elsevier.com/interactiveplots. Interactive Plots provide easy access to the data behind plots. To include one with your article, please prepare a .csv file with your plot data and test it online at http://authortools.elsevier.com/interactiveplots/verification before submission as supplementary material.

The revised version of your submission is due by 08 Mar 2016.

With kind regards,

Roland Olschewski
Editor-in-Chief
Forest Policy and Economics

Is Indonesia utilising its international partners? The driving forces behind Forest Management Units

Muhammad Alif K. Sahide[a], Ahmad Maryudi[b], Supratman Supratman[c], and Lukas Giessen[d]

[a] *Laboratory on Forest Policy and Entrepreneurship, Faculty of Forestry, University of Hasanuddin, Makassar, Indonesia; Chair of Forest and Nature Conservation Policy, University of Göttingen, Germany;*
[b] *Faculty of Forestry, Universitas Gadjah Mada. Bulaksumur, Yogyakarta, Indonesia*
[c] *Laboratory on Forest Policy and Entrepreneurship, Faculty of Forestry, University of Hasanuddin, Indonesia. Makassar, Indonesia*
[d] *Chair of Forest and Nature Conservation Policy, University of Göttingen, Germany.*

ABSTRACT

International forest regimes have been influencing the development of Indonesia's forest policy, and have complemented its domestic policy initiatives. Indonesian political entities utilise the regimes to pursue bureaucratic benefits and national interests. Forest Management Units (FMUs) comprise our heuristic model. We identified international and domestic actors and institutions that underlie the concept of FMUs and how FMUs are implemented. Employing actor-centred analysis, we used observations and expert interviews to distinguish among actors and institutions, as well as various actors' interests in FMU development. We found that the German government via the German company *Deutsche Gesellschaft für Internationale Zusammenarbeit GmbH* (GIZ) is the most powerful actors behind FMUs. British government via the British Department for International Development (DFID), the World Bank and the Asian Development Bank (ADB) have followed in the footsteps of previous actors in terms of directly implementing FMUs. FMUs have a double edge function: to make it easier for international regimes to penetrate Indonesia's domestic politics, and to make central bureaucracies more capable of carrying out programmes and pursuing domestic interests such as reinforcing state forest areas, promoting forest benefits, centralising the budget, capacity building, and centralising information

Key words: Regimes; Interests; International influences; Forest Management Units; Indonesia

1. INTRODUCTION

Forest Management Units (FMUs) or *Kesatuan Pengelolaan Hutan* (KPHs) in the Bahasa Indonesia language, have been a core of domestic forest policy in Indonesia, particularly designated as the main tools for domestic forestry sector's reform. In the Indonesian forest law mandates the implementation of KPH as a prerequisite for SFM in the whole country. FMUs have received broad support from international donors since the 1990s (e.g experimentation of FMU production in South Kalimantan). The implementation of FMUs require high political success on the part of politicians, the capacity and availability of institutions to overcome the problem of the hollow state where a state fails to govern forest management due to an unclear sense of tenure (relating to non-state/community actors' claims to the land) and no clear ownership by various government entities over green zones, as well as a transparent government (Ostrom 1999, Agrawal 2007) in regards to managing forests, and innovative approaches, which is an especially important requirement for implementing the projects of different global regimes at the domestic level.

With a few exceptions such as McDermott et al. (2010), recent research has not widely examined these pathways. Sahide et al. (2015) concluded that international regimes would only be relevant for Indonesia at a high level if domestic actors were politically engaged, even if the regimes did not fit very well with domestic needs and problems. An other example, Indonesian timber certification or *Sistem Verifikasi Legalitas Kayu* (SVLK) has been contested with international regimes by both domestic and international actors (Nurrochmat et al., 2015).

Domestic politicians utilise international regimes to pursue domestic and bureaucratic interests. For example, the Ministry of National Development Planning (BAPPENAS) formally stated that all international donors entering Indonesia should invest their money directly in FMU development, or indirectly steer their program to the location of an FMU. Such a statement can be explained by looking at bureaucratic politics, which show that FMUs not only offer technocratic tools, but also contain many political instruments for actors to pursue domestic interests. MoFor has designated 529 FMU Units or 50% from the target (until 2014), and 120 of them has been designated as FMU Model (until February 2015) (MoFor, 2015a).

This paper will show how international influences have become salient at the domestic level (Berstein and Cashore, 2012), especially in Indonesia (Sahide et al., 2015). In addition, this paper will examine the framework of the pathways that regimes take; domestic actors use this framework (Berstein and Cashore, 2012). This paper supports Gourevitch's (2009) finding that international regimes became an explanatory variable for domestic politics. Overall, using bureaucratic politics theory, this paper views the domestic and international actors and their interests as driving forces underlying the concept of FMUs, and considers how FMUs are implemented in Indonesia. Hence, this paper identifies: the main international and domestic actors and institutions that influence the concept of FMUs and how they are implemented, as well as their interests.

2. THEORETICAL UNDERPINNINGS

2.1. The pathways of influence for international regimes to penetrate domestic politics

International regimes (Humphreys, 1999; Rayner et al., 2010; Edwards and Giessen, 2014) enter the domestic arena through four pathways: (1) direct access, (2) rules, (3) norms, and (4) the market (Bernstein and Cashore, 2012). The traditional goal of international regimes is to influence domestic policies (Bernstein and Cashore, 2012; Sahide et al., 2015; Giessen et al., 2014; Wiersum and Elands, 2013). Bernstein and Cashore (2012) show that actors and the structures of problems (relating to society, politics, and forest issues) determine the forms that regimes take; they also describe how international regimes placed in a national context are different in each country. Global regimes depend on the structures of the aforementioned problems, actors, and the institutional setting.

FIGURE 1. *International regimes influence the notion of FMUs and how they are implemented (adapted from the four pathways framework developed by Bernstein and Cashore, 2012)*

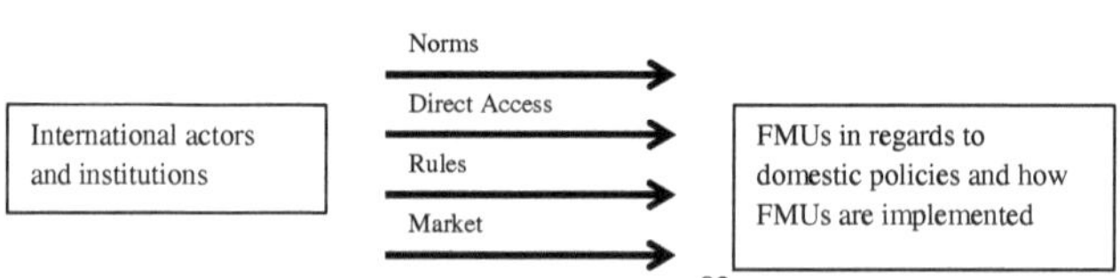

Bernstein and Cashore (2012) developed a framework that differentiates among four distinct pathways through which actors and institutions affect domestic policies: (1) international rules, (2) international norms and discourse, (3) interventions in (or creating) markets, and (4) direct access to domestic policy processes. Based on Figure 1, international regimes could potentially use these four channels to infiltrate FMU political development. In terms of the route for international norms, in order to highlight the issue of regimes at the domestic level, Cortel and Davis (2000) developed four mechanisms: (1) domestic actors materialise their interests, (2) form domestic political institutions, (3) utilise social movements, and (4) develop national campaigns that include political rhetoric.

2.2. Actor-centred analysis: Bureaucratic politics, interests, and power

When we describe actors as a driving force, we are referring to the actor-centred analysis of international and domestic actors as they influence FMU development in Indonesia. Bureaucracies have two central goals: (1) to provide public services using a problem-oriented approach, as stated in their mandates, and (2) to pursue the organisational interests of survival and expansion (Giessen *et al.*, 2014; Giessen, 2011; Krott 1990, 2005; Schusser *et al.*, 2015). From the perspective of bureaucratic politics theory (Niskanen, 1971; Krott, 1990; Peters, 2010), it is logical for bureaucracies to impose their political agendas in opposition to other bureaucracies, especially with respect to who is benefitting the most from the formal structure and informal interests of FMUs.

To measure political influence at the international and domestic levels, we developed a framework in which political influence is a function of information and power (based on Simon, 1981; Krott 1990, 2005). Political influence is the ability to form or implement an international regime's forest issue elements (e.g. international norms of SFM for domestic FMU programmes) according to an individual's or organisation's interests.

Simon (1981) and Krott (1990, 2005) argue that bureaucracies' power mainly stems from: (1) coercing people into accepting their legal mandates (including decision rights and sanction mechanisms); (2) technical resources (enabling bureaucracies to maintain a policy process); (3) alliances between bureaucracies and external actors, who utilise resources to pressure bureaucracies to accept certain preferences; (4) incentives to transfer funds; (5) creating a link to other forest issues; (6) funds that set incentives for others to follow their preferences; (7) bureaucracies' expertise; (8) actively selecting certain information; and (9) the interest-guided use of beliefs and values.

3. Methodology

This paper views that the interests of both domestic and international actors drive the promotion of the concept of FMUs in Indonesia. To shed light on how those actors (and their interests) drive the implementation of FMUs, we employ a number of approaches. Firstly, we use direct observations. In fact, these authors are/were both directly and indirectly involved in implementing FMU projects (which are funded by various international institutions), or were present at some public consultations on international and national projects. This is in line with Van Evera (1997), who maintains that observation is essential for collecting and verifying empirical findings.

Secondly, to accompany the observations, from 2014–2015, we conducted expert interviews with staff members from different government entities related to. Table 1 shows the list of interviews, which we intend to reflect our hypotheses and improve a fundamental clarification for our judgement of the observations (Mosley, 2013). We also conducted a literature review of the content of official documents and news (Hsieh and Shannon 2005; Neuman, 2005) regarding FMU development in Indonesia. The content analysis is based on our expert interviews. We used the most relevant formal documents, as well as text from trusted magazines.

TABLE 1 *List of interviews*

	Interviewee
• **Interview 1, 2015.**	The Asian Development Bank's (ADB) consultant (1) for the Forest Investment Program (FIP) in Indonesia on 1 June 2015
• **Interview 2, 2015.**	The ADB's consultant (2) for the FIP in Indonesia on 2 June 2015
• **Interview 3, 2015.**	Staff from Deutsche Gesellschaft für Internationale Zusammenarbeit GmbH (GIZ) for the Forests and Climate Change Programme (FORCLIME) in Indonesia on 10.06.2015
• **Interview 4, 2015.**	Multistakeholder Forestry Programme (MFP) staff for the MFP 3 (i.e. MFP in its third phase) project in Indonesia
• **Interview 5, 2015.**	Interview with MoFor staff on FMU development in Indonesia (1) on 2 June 2015
• **Interview 6, 2015**	Interview with MoFor staff on FMU development in Indonesia (2) on 2 June 2015
• **Interview 7, 2015.**	Interview with the Regional Director of Management and Areal Preparation on MoFor's use of forest areas, 7 May 2015 (via electronic messages)
• **Interview 8, 2015.**	Interview with the praesidium of the National Forestry Council and the FMU Consultant on 20, 22, & 24 April 2015 (via electronic messages)
• **Interview 9, 2015.**	Interview with the South Sulawesi Provincial Government Officer, who was also the secretary of the FMU establishment team from 2008–2010 (via electronic messages)
• **Interview 10, 2015.**	Interview with the leader of the Indigenous Peoples Alliance of the Archipelago (AMAN) in 2014 (direct interview) and 2015 (via electronic messages)
• **Interview 11, 2015.**	Interview with the secretary of the forest agency of South Sulawesi Province
• **Interview 12, 2014.**	Interview with the Officer of the forest agency of the Bantaeng District Government
• **Interview 13, 2015.**	Interview with MoFor's National Secretary of FMU Implementation
• **Interview 14, 2015.**	Interview with the National Secretary of the Indonesian Communication Forum on Community Forestry (FKKM), which is involved in facilitating the development of Lebak Customary Forest
• **Interview 15, 2015.**	Interview with the staff of Balang Sulsel, which is a non-governmental organisation (NGO) in South Sulawesi Province, and acts as the field facilitator for the management of Kajang Customary Forest, with support from the Center for International Forestry Research (CIFOR) in Bulukumba District, 15 May 2015 (via electronic messages)

4. Results and Discussion

4.1. International driving forces

In the following sections, we detail how international actors support FMUs in Indonesia and how this is inline with their interests

4.1.1. The main international actors behind the concept of FMUs and their implementation

Indonesia's forestry sector has been charaterized by the influence of international actors Sahide and Giessen (2015) and Maryudi et al (2015). In light of Indonesia's decision to reform its forestry sector, we identified four major international actors involved in initiating, developing, and implementing FMUs in Indonesia. They are the German Technical Cooperation Agency (GTZ), the British Department for International Development (DFID), the World Bank, and the ADB (please see Table 2 for more details). GIZ is the strongest and most long-standing actor in terms of introducing FMU development in Indonesia. FMU development was off to a solid start in the early 2000s, and GIZ has been guiding FMU policies ever since in terms of discussions, the policy process, and consultations, as well as making the concept a reality by experimenting with models.

By studying the consistent efforts of GIZ, DFID has also become a pioneering international actor involved in policy and social interventions related to forest governance in Indonesia. DFID brought civil society into Indonesia's forest policy process, including in regards to FMUs. In 2014, DFID began to introduce the purpose of FMUs during MFP 3, and will do so until 2017. MFP 3 integrates the Indonesian forest certification system into FMU policies and development, especially with regards to tenure issues (MFP, 2014). The most recent international programme related to FMU development is the FIP, which is supported by the ADB. FIP was established in 2008 with support from the Climate Investment Funds (CIF) to aid the development and execution of government-led programmes in order to achieve measurable reductions in greenhouse gas emissions and other forest co-benefits (Climate Focus, 2014).

TABLE 2 *Main international actors engage in FMU initiatives and their projects from 2000 – date in Indonesia*

Actors	Project Title	Duration	Financial Volume	Political output (direct and indirectly)
GTZ/GIZ	Strengthening Management Capacity Programme (SMCP)	2000–2008	No data	- Bringing civil society into FMU policy discussions - Solid FMU policies - Established the National Secretary of FMU development
	FORCLIME I	2008–2009	€7.75 million	Experimenting with real FMU models and linking them with global climate change issues
	FORCLIME II	2013–2016	€14.48 million	Experimenting with real FMU models and linking them with global climate change issues
World Bank	FIP	Planning process	WB: $17 million	-
ADB	FIP	2016–2020	ADB: $17 million Indonesia: $1.5 million	- Assist a small number of pilot FMUs to become operational and engage with local communities - Stimulate several MoFor agencies related to FMU implementation
DFID	Improving forest management in Indonesia (MFP 3)	2014–2017	£8.8 million	Linking Indonesia's forest certification system with FMU policies and development, especially in terms of tenure issues

Sources: Interview 3, 2015; Interview 1, 2015; Interview 4, 2015; MoFor, 2015b; ADB and MoFor, 2015; GIZ and MoFor, 2015; MFP, 2014.

4.1.2. Formal motivations to engage in FMU initiatives

Germany is one of the largest donors of development cooperation projects for forestry. The interest of European countries, particularly for Germany can be observed as seeking the professional and logic argumentation to spent their money, which is already allocated for developing and tropical countries like Indonesia. In the formal argumentation, European countries need timber from the sustainable management and good governance system. German Government has clear written mandate of their international forest policy field to effectively assist developing countries on gaining the multiple contributions of forests to the alleviation of poverty, to food security and rural livelihoods as well as to environmental conservation, climate protection and a green economy that need to be more widely acknowledged at all levels (Federal Ministry of Food, Agriculture and Consumer Protection, German Government, 2015; GIZ, 2015). Finally, FMUs' contain several logic reasons to assist international actors in handling technical instruments to infiltrate domestic policy making such as professionalism apparatus, good forest governance, pro poor oriented, multiple purpose of forest production, conflict governance. With this logic, international actors are easily spending their international budget to support FMUs development rather than disbursed into direct domestic regular's activities.

4.1.3. Informal motivations to engage in FMU initiatives

a. Countering global deforestation or countering Indonesian palm oil?

European union (EU), particularly Germany international forest cooperation is one of the main Indonesian international partners on countering global deforestation (Giessen and Sahide, unpublished). Palm oil commodity development on the global discourse is castigated as the one of the mayor factors of deforestation and degradation, particularly in Indonesia (Sahide et al. unpublished). In the informal mission, EU countries realise that palm oil is a big competitor for their own vegetable oils (especially for rapeseed oil). In the global market, high competition over vegetables oil commodities was arisen when a large consumer goods company in the European Union began to shift to the commodity of oil as raw material production. Indonesia as the biggest palm oil producer offer cheap and massive quantity that attract global market, tend to shift the position of western countries in the rivalry, especially in the search for new markets in India and China (Berita Satu, 2014). To block this trend, European campaign 'no palm oil' on their product as the results from deforestation (and health) issues (Taihitu, 2015). Therefore, FMU contains instruments to block deforestation including massive expansion of palm oil. International actors expect FMU can be professionally exist not only to generate income from sustainable forest practice but also as the field agent on reducing deforestation that resulted from palm oil expansion. With FMUs development, illegal palm oil expansion will be effectively reduced, meanwhile FMU is introduced to developed other agricultural commodity which is not compete with palm oil.

b. Maintaining the dominance of international fund providers

The ADB does not want to lose Indonesia as a partner, which is listed as an emerging economic country that can pay back the loan for their projects including FIP. At the Asia and Africa Summit, held in April 2015, Indonesia's president argued that developing countries should no longer depend solely on the World Bank and the IMF; rather, he asserted that Indonesia should engage with developing countries to find alternative global financial institutions (Antara, 2015). Indonesia is now aligning with China to create a new financial entity: the Asian Infrastructure Investment Bank (AIIB). The Indonesian government had to sign the Article of Agreement (AoA) that establishes this financial group (Stelindonesia News, 2015). This will give a pressure for ADB, World Bank, and IMF due to the new exist of global competitor fund provider as well as give Indonesia some leverage in bargaining its interests to define the project's underlying ideas (See section 4.2.1 for the domestic actor's interests).

4.1.4. Pathways used by international actors influencing FMUs

The discourse on international norms of sustainable forest management (SFM) and global direct access programs significantly influences FMUs. Through direct access and rules (Bernstein and Cashore, 2012), the World Bank and the International Monetary Fund (IMF) have a very big influence on the economic development resulting from Indonesia's transformation policies and activities related to tropical forests (Sahide et al., 2015). During the Indonesian financial crisis of 1999, Indonesia received IMF funds and restructured its economy, including the forestry sector. During the timber boom era in 1980s, the World Bank and the IMF were the most important international actors behind domestic projects (Sahide and Giessen, 2015). This is in line with Gourevitch's (2009) finding that international market forces affect politics. Recently, Indonesia has witnessed massive deforestation, and conservation-oriented international regimes are now entering Indonesia via methods of international norms discourse such as SFM, which brought diverse regimes to the country. Table 3 displays the international actors and institutions that introduced SFM norm and discourse via FMUs.

The pathways of the market and rules are no longer relevant for FMU issues directly. However, norms have become a more salient strategy in domestic politics because international actors through direct access pathway make it relevant. SFM discourse, was adjusted to fit the Indonesian context through technocratic consultancy which is provided by international actors under direct access pathway. In terms of technocratic consultancy, ideas such as state professionalism, decentralisation, pro poor, transparency, and public participation drive the normative SFM discussions into Indonesian FMUs policy.

4.2. Domestic driving forces

In the following sections, we analyse how the main domestic actors use the FMU concept in pursuing their interests

4.2.1. FMU's for attracting international support: preparing direct access

The notion of FMUs derived from technocratic expertise, which involves both national and international experts. Technocratic expertise altered Indonesia's legal forestry governance system, which consists of two parts; Sahide and Giessen (2015) concluded that it is a very complex, fragmented structure. Technocratic thinking has produced intricate FMU instruments (such as mechanisms for regulations, budget incentives, information, and responsibility), which are very useful for international regimes. These instruments are created by the regulatory powers of the many central bureaucracies involved. Table 3 shows how Indonesian bureaucracies form FMU policies to match with international regimes' requirements for conducting their activities in Indonesia.

TABLE 3 *The FMU instruments that domestic actors use to attract international regimes*

Actors that create the instrument	FMU instrument	The international regimes that used the instrument, and related actors	Purpose
1. Ministry responsible for forestry (MoFor)	1. Formal forest area appointed	1. Climate change and Reducing Emissions from Deforestation and Forest Degradation (REDD+): the United Nations Framework Convention on Climate Change (UNFCC)	1. Clarify areas for transparent monitoring,
2. Ministry of Home Affairs (MoHA)	2. Forest institution at the forest site	2. Biological diversity: Convention on Biological Diversity (CBD)	2. Clear the process for checking and assessing the supply of timber in terms of the certification regime
3. Ministry of State Apparatus and Bureaucracy Reform (KemPAN)	3. Formal responsibility arrangement between central government and provincial/district governments	3. Trade in endangered species: Convention on International Trade in Endangered Species (CITES)	3. Clear forest areas to conduct a biodiversity inventory
4. BAPPENAS	4. The FMU institution is monitoring forest permits (community forest schemes and timber concessions)		4. Designate the state forest area as a restricted zone for palm oil plantation

	5. The administration authority is monitoring non-forest permits (mining concessions) 6. Multiple possibilities for the state budget strategy from the central and regional governments (i.e. the provincial and district governments)	4. Wetland management, 5. International tropical trade: the International Tropical Timber Organization (ITTO) 6. Illegal logging and timber certification: the Forest Stewardship Council (FSC), the Programme for the Endorsement of Forest Certification (PEFC) 7. Palm oil certification: the Roundtable on Sustainable Palm Oil (RSPO) 8. Forestry research regime: CIFOR 9. The Association of Southeast Asian Nations (ASEAN) forest and environmental regime: ASEAN Senior Officials in Foresty (ASOF) 10. International forum for sharing information on forestry regimes: the United Nations Forum on Forests (UNFF)	5. Forest permits are used for to safeguard an area for some related regimes 6. Identifying illegal uses of state forest areas and forest resources 7. State representative for conflict resolution, based on institutions in the field 8. State representative for developing agroforestry and community forestry in the field

Table 3 shows that FMU instruments in Indonesia include high political will, institutional capacity, and innovative approaches to address the driving forces behind deforestation. Deforestation is a particularly important criterion for developing and implementing different international regimes that deal with diverse topics, particularly REDD+. REDD+ match with FMUs mission, who is seeking more concrete instruments to fill many gaps in domestic implementation, such as a monitoring system, a benefit sharing mechanism, and other technical aspects at the domestic level (Kim et al. 2015; Brockhaus et al., 2012). The domestic actor invites international regimes by using the argument that KPH fits with REDD

Tables 3 and 4 present the details of international projects that successfully attracted direct and indirect support for FMUs. We observed issues such as community forestry, small-scale forestry businesses, timber legality, and illegal logging; FMUs can cover all these topics. Climate change is the most consistent and attractive domestic matter supported by the multiple international actors mentioned in this paper.

Hence, domestic actors such as MoFor, the National Land Administration System (NLAS), BAPPENAS, MoHA, and KemPAN are the driving forces behind strong FMU development, which has successfully attracted diverse international regimes.

MoFor's new structure is one of the central bureaucratic frameworks that strongly supports FMU development. MoFor created positions for 11 directorates general (DGs) and 4 of them are responsible for FMU development. In lower level, four directorate institutions (which DGs run) formed that are responsible for FMU development (Forda, MoFor, 2015). Furthermore, domestic actors have successfully used FMUs to recentralise all power back to MoFor. For example, BAPPENAS declared that '[without] FMUs, [there would be] no budget' and launched a campaign with the slogan: 'no FMUs, no aid' (Setyarso, 2014). BAPPENAS has formally stated that all international donors that invest in Indonesia should directly place their money in FMU development or indirectly situate their programme at an FMU site or planning area.

TABLE 4 *International projects guided at FMU locations or indirectly related to FMU development*

Actor	Project	Period	Budget
International Forest Carbon Initiative (IFCI, Australia)	Kalimantan Forest Climate Partnership (KFCP)	2007/08–2011/12	~$30 million
IFCI (Australia)	The Indonesian National Carbon Accounting System (INCAS)	2007/08–2011/12	~$2 million
The Australian Centre for International Agricultural Research (ACIAR)	Improving governance, the policy, and institutional arrangements to Reduce Emissions from Deforestation and Degradation (REDD)	2008/09–2011/12	~$1.4 million
The Norwegian government	Norway's International Climate and Forest Initiative (NICFI)	2010–2016 (or longer)	$1 billion
The Norwegian government	UN-REDD Programme	2010–2011 (phase 1)	$5.6 million
The World Bank	Forest Carbon Partnership Facility (FCPF)	2011–2013	$3.6 million
The EU	REDD+	2011	€6 million
KfW Development Bank (Germany's Federal Ministry for Economic Cooperation and Development, or BMZ)	Forest Programme (Support for MoFor)	2010–2017	$27.5 million
KfW (Germany's Federal Ministry for the Environment, Nature Conservation and Nuclear Safety, or BMU)	Pilot Project: Ecosystem Restoration in the Harapan Rainforest	2009–2013	€10.4 million
United Nations Office on Drugs and Crime (UNODC)	Strengthening Law Enforcement Capacity in Countering Forest Crimes when implementing REDD+	2011–2012	$1.29 million
Swiss State Secretariat for Economic Affairs (SECO)	Support from the presidential task force to oversee REDD+ activities, including coordination of related financial resources from international donors	July 2012-July 2014	$900,000

Source: MoFor, 2015; Interview 5, 2015; Interview 6, 2015.

Indonesia attempted to negotiate with the ADB to steer the technical implementation of ADB's FIP. For example, Indonesia's government negotiate the strict interpretation of ADB guidelines that do not fit with Indonesia's legal and domestic contexts. ADB did not want to pay for FMUs because they are government organisations, which is contrary to ADB's guidelines. The Indonesian consultants (who come from NGOs and academic settings) thought that FMUs should not be funded since they could make a profit, as their areas are large and still at the experimental stage.

One of MoFor's key officials argued that Indonesia could reject ADB's FIP (which is still in operation today) related to FMUs if it did not suit Indonesia's priority system (which is part of the legal system). Indonesia's firm attitude about overseeing the FMU project resulted from its increasingly strong economic position in terms of its gross domestic product (GDP).

4.2.2. MoFor reinforcing state forest areas

In terms of Indonesia's land system, land is divided into two categories: state forest areas and land outside forest areas. These zones fall under the jurisdiction of the state; in this case, provincial and district governments (Bakker & Moniaga, 2010), but their status is interpreted in diverse ways, especially concerning the legal definition of a 'state' forest area (Sahide and

Giessen, 2015). Sixty per cent of all land in Indonesia consists of state forest areas administered by MoFor. Following the 'big bang' decentralisation at the end of the 1990s, regional governments were granted the authority to supervise production and protection forests, including the power to issue forest management rights to private actors (Sahide et al., unpublished). District governments do not have enough officers on staff, and a clear system of forest management is lacking, which leads to people treating state forest areas as public resources (Sahide et al. unpublished). The overall impact of the district government's fragmented, decentralised power has resulted in massive deforestation, which peaked around 2000 and 2001 (Sahide and Giessen, 2015).

Experimentation with FMUs is highly successful at the political level in terms of recovering control of state forest areas and filling the gap in 'hollow' state forest management. Following MoFor's efforts to restore its authority and gain complete control of timber licences (Tacconi et al., 2014), a new law on regional governments (Law 23 of 2014) was issued to reduce decentralisation by shifting authority away from district governments to provincial ones (please see the sub-section in this paper called 'Central bureaucracies counter decentralisation'). Given this regulatory power, FMUs are now used to enhance MoFor's position to recover state forest areas.

4.2.3. MoFor re-centralising project budget

Recently, MoFor evaluated the previous international FMU project, particularly in terms of transferring the FMU's budget to the stages of implementation. GIZ programmes, such as FORCLIME, and DFID activities (such as policy advice and capacity building) are funded with grants from the BMZ or the British government. These grants are directly given to GIZ and DFID and are not distributed to other partners (of BMZ and the British government); thus, the grants are 'part of the budget but off the treasury record'. The grants are used for various purposes including operational costs, staff expenses, consultancies, workshops, trainings, and publications. In addition, FORCLIME, which is run by MoFor, received a grant from the BMZ; this grant was integrated into MoFor's budget and distributed to the forest district administrations in pilot districts.

MoFor agrees with the 'part of the budget but off the treasury record' mechanism, and evaluates the transfer of a budget from its own offices to a district government's budget. From MoFor's perspective, it is ineffective to use the entire amount of the budget due to some complicated problems with disbursing money via district bureaucracies. To avoid this challenge during the next project negotiation, especially for the FIP on FMUs, MoFor asked the ADB for an FIP through which the money would not be transferred to the district government, but rather to related MoFor agencies in the central and regional governments. MoFor now has more flexibility to advocate for the budget provided and fully implement projects in the field. The budget transfer mechanism is replicated through MoFor's financial control, which is based on the National Forest Rehabilitation Programme (GERHAN). GERHAN is a MoFor agency that distributes budgets to various users such as the local government, consultants, and private actors.

4.2.4. MoFor centralising capacity building

GIZ also actively works with MoFor's Training Centre Agency to develop training curricula for FMU capacity training programmes in selected national parks that focus on FMU production or conservation forests (FORCLIME, 2015). In addition, GIZ is also involved in increasing the capacity of MoFor's Forestry Senior High School to manage

FMUs (KPH Dephut, 2015). MoFor has allocated a budget to hire regular forestry students, as well as students studying for their bachelor's degree in forestry, as field staff for selected FMU models. Capacity building interventions and programmes are useful for MoFor to demonstrate its control of FMU development programmes. Several training activities related to FMU programmes have been conducted by the Training Centre Agency with GIZ support at certain times and for specific purposes. MoFor officials often complain that local government staff train in FMUs, and then suddenly shift to sectors not associated with FMU programmes. This indicates that FMUs have somehow not become a priority for local governments. Recently, MoFor developed a certification system (which is currently being tested in a pilot phase) through which people eligible to manage FMUs can become certified. This system would be mandatory for all FMU field staff. Thus, MoFor has strengthened its position as a hub of capacity building for all local governments that want to build FMUs in their jurisdictions.

4.2.5. MoFor centralising information

Information is a very powerful tool that can propel and reduce the user's biased interests (Krott, 2005; Wibowo and Giessen, 2015). MoFor controls key information about forests and FMUs such as state and non-state forest locations, changing the status of state forest areas to non-state forest areas, the baseline details on forest conditions, international projects in state forests, legal instruments to develop FMUs, and the availability of FMU experts. Not only can MoFor use this essential knowledge to manipulate users' awareness of FMUs; it can also block the biased interests of local governments and private actors that are contrary to MoFor's interests.

Different versions of the FMU concept and the diverse challenges of implementing FMUs result from actors' varied interests. To combat this problem, MoFor should manage information via strong bureaucracies, as well as events supported by a task force (whose members are appointed by MoFor). The national secretary of FMUs (Setnas, KPH) was a position created by MoFor (through MoFor Decree SK.511/Menhut-II/2012), and also supported by GIZ (MoFor, 2012; KPH Dephut, 2012). This position aids multiple actors in developing the FMU concept and monitoring the implementation of FMUs; in addition, the national secretary acts as a hub of information and technical assistance (MoFor 2012).

4.2.6. Central bureaucracies countering decentralisation

Currently, there are at least two different ideas to make FMUs into institutions: by declaring that a forest (which is under FMU supervision) exists for the purpose of (1) production or (2) protection (Sahide et al., unpublished). The first proposal is that the staff of a specific FMU should learn to administer the ecological watershed area, based on the watershed approach. Provincial governments mostly use this strategy to establish FMUs in inter-district areas. The second idea is that the second type of FMU should employ the basic principle of forest management, which is that units should serve the administrative body at the specific forest site, reflecting the purpose of the forest (i.e. production or protection). District governments mostly use this argument to establish FMUs under their jurisdiction.

Law 23 of 2014, a new regional government law, mandates that the forestry sector be recentralised to the authority of the provincial governments, based on the previous situation of decentralising district governments (Sahide et al., unpublished). It is easier to handle the interests of 34 provincial governments, rather than a very broad range of interests from more than 500 district and municipal governments. The new situation has legal support (Law 23 of

2014) in that provincial governments are the representatives of the central government in regional areas. Following the huge amount of deforestation that has resulted from uncontrolled permits for plantations issued by the district government, MoFor has brought some forest regulations back to the provincial and central governments (Tacconi et al., 2014; Bae et al., 2014). This is an ideal moment for MoFor to campaign and establish FMUs as tools for recentralising and countering a complicated situation of forestry decentralisation (Sahide et al., unpublished). This is in line with Ribot et al. (2006) and Mutebi (2004), who showed that decentralisation policies could be a strategy for recentralisation.

However, these efforts face two challenges from district governments that want to continue to administer forest areas (Sahide et al., unpublished); those two challenges are (1) administering state forest areas and (2) administering non-state forest areas. Furthermore, indigenous actors launched a campaign slogan: 'no rights, no FMUs' to counter the massive goal of FMUs to re-affirm the legal status of state forests. This means that MoFor has to guarantee indigenous people's rights before implementing FMUs.

4.2.7. Central bureaucracies promote forest benefits: Countering conservation

In a context where conservation is a big concern for international regimes, global timber certification and other conservation-oriented regimes are now the big global actors influencing domestic policies, such as making timber certification mandatory in Indonesia (Wibowo et al., unpublished). The concept of FMUs could be used to facilitate conservation-oriented regimes with global interests, as well as domestic, production-oriented regimes. For example, under the MFP 3 project, FMUs could potentially link up with global and domestic timber certification regimes. International norms on SFM could be adopted by experimenting with FMUs; SFM is now causing a famous slogan to spread within Indonesia: 'sustainable forests equal people's welfare'. Since the forestry sector is now shaping a legal mandate within a framework of green investment, MoFor is trying to not only conserve forest areas, but also generate income from them.

International actors will be actively involved in developing a road map for revitalising Indonesia's forestry industry until 2020. MoFor and BAPPENAS conducted a study involving the international actors, which they synthesised to develop recommendations for creating a policy on Indonesia's forestry industry (Justianto and Sukmananto, 2005). These actors and their activities included: (1) Strategies for the Development of Sustainable Wood-based Industries by ITTO; (2) An Analysis of Future Forest Scenarios by a partnership between United States Agency for International Development (USAID) Natural Resource Management and DFID's MFP; and (3) Generating Economic Growth, Rural Livelihoods, and Environmental Benefits by CIFOR and the World Bank. In the study's conclusion, MoFor and BAPPENAS recommended that efforts be made to connect the forestry industry with SFM programmes (which include FMU development), and to overcome illegal logging activities. Hence, these results indicate that FMU development is one of MoFor's strategies.

In terms of FMU conservation, current experiments are happening whereby national parks are re-named as FMU conservation forests; the status quo of national park management under the supervision of the central government continues. This shows that MoFor has very much served FMU production and protection forests, rather than FMU conservation forests. In order to counter the legal premise that conservation areas are the most restricted zones for community activities, many international actors have established their projects in national parks (i.e. FMU conservation forests).

5. Conclusion

FMUs have a dual edge function: (1) to assist international actors in handling technical instruments so they can infiltrate domestic policy making and implementing their political agenda (e.g. finding professional reasons to spent international cooperation budget, countering global deforestation, and maintaining the established and dominance international fund provider); and (2) to serve the interests of domestic actors mainly through getting international financial support and coalition partner, intensify forest utilization, recovering back the state forest area, as well as recentralize competencies, information and funds.

We found that the German government are the most powerful actors behind FMUs. Germany impacts FMUs via the German company *Deutsche Gesellschaft für Internationale Zusammenarbeit GmbH* (GIZ), which specialises in international development. British government via via the British Department for International Development (DFID), the World Bank and the Asian Development Bank (ADB) have followed in the footsteps of previous actors in terms of directly implementing FMUs. International actors materialise domestic actors' interests, and socialise political forces through domestic regulatory power. Meanwhile, BAPPENAS, MoHA, and KemPAN aligned with MoFor to make FMU programmes a national interest. The results show that all components of power (coercion, incentives and disincentives, and controlled information) have been successfully recentralised back to MoFor by experimenting with the FMU concept and FMU implementation via programmes supported by the international actors. MoFor successfully utilised FMUs to recover state forest areas, centralise information and the budget mechanism, and use decentralised FMU policies for recentralisation.

Multiple central bureaucracies support FMUs, which became a national interest. FMU development contains a high amount of political success that attracted international actors and institutions to link their programmes with FMUs. MoFor actively promotes sustainable and economic use of forests, as opposed to conservation instruments, which are often seen via their 'technical' characteristics. This means that an instrument solves a political problem at best. For example, consider climate change, a problem for which economic instruments such as taxes or tradable permits are seen as being technically better than regulatory tools.

FMU instruments have shown a strong tendency toward political success. FMUs are not only utilised due to their technical abilities, but also to reflect certain political interests. Many instruments are not introduced due to their 'technical' features with regard to problem solving. Indonesia used a regulatory instrument (which is better in this context than another tool such as a tax or economic instrument); this reflects the government's political interests more clearly.

We also identified three arenas of potential conflict of interests, such as 1) among actors between international and domestic actors in countering deforestation or promoting benefits from forests, 2) international actors and other international actors on imposing internal agenda, and 3) among domestic bureaucracies between central, province and district government in manifesting FMU as decentralisation policy or recentralisation strategy (Sahide et al., unpublished)

ACKNOWLEDGMENT

We greatly appreciate the financial support from the research project of Deutsche Forschungsgemeinschaft (DFG) on the fragmentation of the international forest regime complex (PAK 813), the Eva Mayr-Stihl Foundation

REFERENCES

Bae, J. S., Kim, Y. S., Fisher, L., Moeliono, M., & Deshazo, J. 2014. Promises and Perils of Decentralized Forest Governance: The Case of Indonesia's Forest Management Units in Reducing Emission from Deforestation and Forest Degradation (REDD+). *Society & Natural Resources*, **27**(12), 1346-1354.

Bakker, L., Moniaga, S., 2010. The Space Between: Land Claims and the Law in Indonesia. Asian Journal of Social Science 38 (2), pp. 187–203. DOI: 10.1163/156853110X490890.

Bernstein, S., Cashore, B., 2012. Complex global governance and domestic policies: four pathways of influence. Int. Affairs 88 (3), 585–604. DOI: 10.1111/j.1468-2346.2012.01090.x

Brockhaus, M., Obidzinski, K., Dermawan, A., Laumonier, Y., & Luttrell, C. (2012). An overview of forest and land allocation policies in Indonesia: Is the current framework sufficient to meet the needs of REDD+?. Forest policy and economics, 18, 30-37. http://dx.doi.org/10.1016/j.forpol.2011.09.004

Cortell, A. P., & Davis Jr, J. W. (2000). Understanding the domestic impact of international norms: A research agenda. International Studies Review, 2(1), 65-87.

Edwards, P., Giessen, L., 2014. Global forest governance — Discussing legal scholarship from political science perspectives. Forest Policy and Economics, 38 (2014), 30-31, http://dx.doi.org/10.1016/j.forpol.2013.08.006

Giessen, L., Krott, M., Möllmann, T. 2014. Increasing representation of states by utilitarian as compared to environmental bureaucracies in international forest and forest-environmental policy negotiations. In: Forest Policy and Economics, 38, 97–104. doi:10.1016/j.forpol.2013.08.008

Giessen, L. 2011. Reviewing empirical explanations of policy change: Options for its analysis and future fields of research. In: *Allgemeine Forst- und Jagdzeitung*, **182** (11-12), 248-259

Giessen, L. Sahide, and M. A. K. unpublished. Blocking, Attracting, Imposing, and Allying: The utility of ASEAN for member states in forest and environmental politics

Gourevitch, P. 1978. The second image reversed: the international sources of domestic politics. International Organization, 32(04), 881-912. DOI: http://dx.doi.org/10.1017/S002081830003201X

Hsieh, H.F., Shannon, S.E. 2005. Three approaches to qualitative content analysis. Qual. Health Res. 15, 1277–1288. DOI: 10.1177/1049732305276687

Humphreys D. 1999. The evolving forests regime. Global Environmental Change 9(3): 251–254. http://dx.doi.org/10.1016/S0959-3780(99)00005-9

Justianto, A., Sukmananto, B. 2005. Revitalisasi Industri Kehutanan: Permasalahan dan Penanganannya. Agrimedia, 10(2), 19-25.

Kim, Y. S., Bae, J. S., Fisher, L. A., Latifah, S., Afifi, M., Lee, S. M., & Kim, I. A. (2015). Indonesia's Forest Management Units: Effective intermediaries in REDD+ implementation?. Forest Policy and Economics. In Press doi:10.1016/j.forpol.2015.09.004

Krott, M. 1990. Öffentliche Verwaltung im Umweltschutz. Ergebnisse einer behördenorientierten Policy-Analyse am Beispiel Waldschutz, W. Braumüller Verlag, Wien

Krott, M. 2005. Forest policy analysis. Springer, Dordrecht. 323pp.

Krott, M., Bader, A., Schusser, C., Devkota, R., Maryudi, A., Giessen, L., & Aurenhammer, H. 2014. Actor-centred power: The driving force in decentralised community based forest governance. Forest Policy and Economics, 49, 34-42. doi:10.1016/j.forpol.2013.04.012

Maryudi, A., Citraningtyas, E. R., Purwanto, R. H., Sadono, R., Suryanto, P., Riyanto, S., & Siswoko, B. D. (2015). The emerging power of peasant farmers in the tenurial conflicts over the uses of state forestland in Central Java, Indonesia. Forest Policy and Economics, In Press. doi:10.1016/j.forpol.2015.09.005

McDermott, C.L., Cashore, B., Kanowski, P., 2010. Global environmental forest policies: An international comparison. Earthscan, London, Washington, DC.

Mosley, L., 2013. Introduction. just talk to people? interviews in contemporary political science. In: MOSLEY, L. (Ed.), Interview Research in Political Science. Cornell University Press, p. 2013.

Niskanen, W. 1971. Bureaucracy and Representative Government. Aldine- Atherton, Chicago

Nurrochmat, D. R., Dharmawan, A. H., Obidzinski, K., Dermawan, A., & Erbaugh, J. T. 2014. Contesting national and international forest regimes: Case of timber legality certification for community forests in Central Java, Indonesia. Forest Policy and Economics. In Press doi:10.1016/j.forpol.2014.09.008

Peters, B.G. 2010.The Politics of Bureaucracy- An Introduction to Comparative Public Administration. Sixth edition.Routledge. Oxon

Rayner, J., Humphreys, D., Perron Welch, Frederic, Prabhu, Ravi, Verkooijen, P., 2010. Introduction. In: Rayner, J., Buck, A., Katila, P. (Eds.), Embracing Complexity: Meeting the Challenges of International Forest Governance.vol.28. IUFRO World Series,Vienna, pp. 9–18.

Ribot, J. C., Agrawal, A., & Larson, A. M. 2006. Recentralizing while decentralizing: how national governments reappropriate forest resources. World development, 34(11), 1864-1886. doi:10.1016/j.worlddev.2005.11.020

Sahide, M. A. K., and Giessen, L. 2015. The fragmented land use administration in Indonesia–Analysing bureaucratic responsibilities influencing tropical rainforest transformation systems. Land Use Policy, 43, 96-110. doi:10.1016/j.landusepol.2014.11.005

Sahide, M. A. K., Nurrohmat, D.R. and Giessen, L. 2015. The regime complex for tropical rainforest transformation: analysing the relevance of multiple global and regional land use regimes in Indonesia. Land Use Policy, 47, 408-425. http://dx.doi.org/10.1016/j.landusepol.2015.04.030

Sahide, M.A.K., Supratman, S., Maryudi, A., Giessen, L. Unpublished. Decentralisation Policy as Recentralisation Strategy: Forest Management Units and Community Forestry in Indonesia. Unpublished

Schusser, C., Krott, M., Movuh, M. C. Y., Logmani, J., Devkota, R. R., Maryudi, A., Salla, M., And Bach, N. D. 2015. Powerful stakeholders as drivers of community forestry—Results of an international study. Forest Policy and Economics, 58, 92-101. doi:10.1016/j.forpol.2015.05.011

Sikor, T. and Lund, C. 2009. Access and property: a question of power and authority. Development and Change, 40(1), 1-22.

Simon, H.A.1981. *Entscheidungsverhalten in Organisation*. Deutsche Übersetzung, 3. Auflage, Landsberg.

Tacconi, L., Obidzinski, K., and Agung, F. 2004. *Learning lessons to promote forest certification and control illegal logging in Indonesia*. CIFOR.

Van Evera, S., 1997. *Guide to Methods for Students of Political Science*. Cornell University Press.

Wibowo, A., & Giessen, L. 2015. Absolute and relative power gains among state agencies in forest-related land use politics: The Ministry of Forestry and its competitors in the REDD+ Programme and the One Map Policy in Indonesia. Land Use Policy, 49, 131-141. doi:10.1016/j.landusepol.2015.07.018

Wibowo, A., Sahide, M. A. K., & Giessen, L. unpublished. From voluntary private to mandatory state governance in Indonesian forest certification: Reclaiming authority by bureaucracies.

Wiersum, K. F., & Elands, B. H. (2013). Opinions on legality principles considered in the FLEGT/VPA policy in Ghana and Indonesia. Forest policy and economics, 32, 14-22. doi:10.1016/j.forpol.2012.08.004

Empirical Material

ADB and MoFor, 2015. Program Investasi Kehutanan Investasi Khusus Berbasis Masyarakat Untuk Mengatasi Deforestasi dan Degradasi Hutan.

Antara, 2015. Jokowi`s speech one major leap for Indonesia. http://www.antaranews.com/en/news/98660/jokowis-speech-one-major-leap-for-indonesia Accessed on 20 September 2015

Berita Satu, 2014. *Melawan Kampanye Hitam Terhadap Kelapa Sawit Indonesia.* http://www.beritasatu.com/ekonomi/194831-melawan-kampanye-hitam-terhadap-kelapa-sawit-indonesia.html Accessed on 15 October 2015

Climate Focus, 2014. Linkages Between REDD+ Readiness and The Forest Investment Program. http://www.climatefocus.com/sites/default/files/Linkages_between_REDD_readiness_and_FIP_Nov2014.pdf Accessed on 20 September 2015

Federal Ministry of Food, Agriculture and Consumer Protection, German Government, 2015. German forests. Nature and economic factor. http://www.bmel.de/SharedDocs/Downloads/EN/Publications/GermanForests.pdf?__blob=publicationFile Accessed on 15 October 2015

FORCLIME, 2015. http://www.forclime.org/id/18-bahasa-indonesia/berita Accessed on 20 September 2015

FORDA MoFor, 2015. http://www.forda-mof.org//files/Lampiran_P18_2015_Struktur_Organisasi_KLHK.pdf Accessed on 20 September 2015

GIZ, 2015. http://www.giz.de/en/html/about_giz.html Accessed on 15 October 2015

GIZ and MoFor, 2015. Forests and Climate Change Programme Laporan Programme Progress Review, Fase Pertama 2009-2010 dan Perencanaan Fase Kedua 2013-2016.

Kartodihardjo, H., Nugroho, B., Putro, H.R., 2011. Forest Management Unit Development, (FMU) Concept, Legislation and Implementation . GIZ and MoFor

KPH DEPHUT, 2015. http://www.kph.dephut.go.id/index.php?option=com_content&view=article&id=431%3Alokakarya-pengembangan-smk-kehutanan-dukung-pembangunan-kph&catid=1%3Aberita-kph&Itemid=165 Accessed on 20 September 2015

KPH DEPHUT, 2012. *Rapat Pembahasan Sekretariat Nasional Pembangunan Kesatuan Pengelolaan Hutan (KPH)* http://kph.dephut.go.id/index.php?option=com_content&view=article&id=75%3Arapat-pembahasan-sekretariat-nasional-pembangunan-kesatuan-pengelolaan-hutan-kph&catid=1%3Aberita&Itemid=165 Accessed on 20 September 2015

Law 23 of 2014 on Regional Government. *Undang-Undang Nomor 23 tentang Pemerintahan Daerah*

MFP, 2014. Akses Masyarakat ke Hutan. http://www.mfp.or.id/index.php/id/akses-masyarakat-ke-hutan Accessed on 22 September 2015

MoFor, 2012. *Pembentukan Sekretariat Nasional Pembangunan KPH.* http://kph.dephut.go.id/index.php?option=com_content&view=article&id=159:pembentukan-sekretariat-nasional-pembangunan-kph-&catid=1:berita-kph Accessed on 20 September 2015

MoFor, 2015a. Kebijakan dan "Mainstreaming" Pembangunan Kesatuan Pengelolaan Hutan (KPH). Direktorat Wilayah Pengelolaan Hutan dan Penyiapan Areal Pemanfaatan Kawasan Hutan

MoFor, 2015b. Daftar kegiatan kerjasama kementerian kehutanan.

Stelindonesia News, 2015. *Desember, Bank Infrastruktur Asia Beroperasi.* http://steelindonesianews.com/detail.asp?id=1734 Accessed on 20 September 2015

Taihitu, B.P., 2015. *Sentimen Negatif Sawit Dimanfatkan Produsen Makanan*. SAWIT INDONESIA Edisi Agustus-September 2015

Muhammad Alif Kaimuddin Sahide
31.08.1979
Rappang, South Sulawesi
Married, 2 Children

Theodor-Heuss-Straße 11, room 413
37075 Göttingen
Germany
msahide@gwdg.de

Curriculum vitae

Muhammad Alif Kaimuddin Sahide

Formal and Scientific Education

1998-2003 : S. Hut (Bachelor in Forestry), Hasanuddin University, Makassar-Indonesia

2004-2006 : M.Si. (Master of Science in Resource Economics), Hasanuddin University, Makassar, Indonesia

2012-2016 : PhD in Forest Policy – Georg August University Goettingen, Germany Thesis: 'Utilizing international land use regimes to shape domestic forest policies in Indonesia'

Permanent Position

Lecturer in Forest Policy, Forest Economics and Community Forestry at Faculty of Forestry, University of Hasanuddin, Indonesia (Since 2008).

Scholarships/Grants

Indonesian Government Scholarship (DIKTI) PhD study (2012-2016)